K-12 STEM Education in Urban Learning Environments

Jillian L. Wendt
University of the District of Columbia, USA

Danielle L. Apugo
University of the District of Columbia, USA

A volume in the Advances in Early Childhood and K-12 Education (AECKE) Book Series

IGI Global
DISSEMINATOR OF KNOWLEDGE

Published in the United States of America by
IGI Global
Information Science Reference (an imprint of IGI Global)
701 E. Chocolate Avenue
Hershey PA, USA 17033
Tel: 717-533-8845
Fax: 717-533-8661
E-mail: cust@igi-global.com
Web site: http://www.igi-global.com

Library of Congress Cataloging-in-Publication Data

Names: Wendt, Jillian L., 1979- editor. | Apugo, Danielle L., 1987- editor.
Title: K-12 STEM education in urban learning environments / Jillian L. Wendt and Danielle L. Apugo, Editors.
Other titles: K-12 science, technology, engineering, mathematics education in urban learning environments
Description: Hershey, PA : Information Science Reference, [2019] | Includes bibliographical references.
Identifiers: LCCN 2018036349| ISBN 9781522578147 (hardcover) | ISBN 9781522578154 (ebook)
Subjects: LCSH: Education, Urban. | Science--Study and teaching. | Technology--Study and teaching. | Engineering--Study and teaching. | Mathematics--Study and teaching. | Educational equalization.
Classification: LCC LC5129 .K25 2019 | DDC 372.35/044--dc23 LC record available at https://lccn.loc.gov/2018036349

This book is published in the IGI Global book series Advances in Early Childhood and K-12 Education (AECKE) (ISSN: 2329-5929; eISSN: 2329-5937)

British Cataloguing in Publication Data
A Cataloguing in Publication record for this book is available from the British Library.

For electronic access to this publication, please contact: eresources@igi-global.com.

Advances in Early Childhood and K-12 Education (AECKE) Book Series

ISSN:2329-5929
EISSN:2329-5937

Editor-in-Chief: Jared Keengwe, University of North Dakota, USA

MISSION

Early childhood and K-12 education is always evolving as new methods and tools are developed through which to shape the minds of today's youth. Globally, educational approaches vary allowing for new discussions on the best methods to not only educate, but also measure and analyze the learning process as well as an individual's intellectual development. New research in these fields is necessary to improve the current state of education and ensure that future generations are presented with quality learning opportunities.

The **Advances in Early Childhood and K-12 Education (AECKE)** series aims to present the latest research on trends, pedagogies, tools, and methodologies regarding all facets of early childhood and K-12 education.

COVERAGE

- K-12 Education
- Standardized Testing
- Individualized Education
- Reading and Writing
- Performance Assessment
- Special Education
- Head Start and Pre-K Programs
- Learning Outcomes
- Urban K-12 Education
- Curriculum Development

IGI Global is currently accepting manuscripts for publication within this series. To submit a proposal for a volume in this series, please contact our Acquisition Editors at Acquisitions@igi-global.com or visit: http://www.igi-global.com/publish/.

The Advances in Early Childhood and K-12 Education (AECKE) Book Series (ISSN 2329-5929) is published by IGI Global, 701 E. Chocolate Avenue, Hershey, PA 17033-1240, USA, www.igi-global.com. This series is composed of titles available for purchase individually; each title is edited to be contextually exclusive from any other title within the series. For pricing and ordering information please visit http://www.igi-global.com/book-series/advances-early-childhood-education/76699. Postmaster: Send all address changes to above address.

Titles in this Series

For a list of additional titles in this series, please visit:
https://www.igi-global.com/book-series/advances-early-childhood-education/76699

Advanced Strategies and Models for Integrating RTI in Secondary Schools
Pam L. Epler (Grand Canyon University, USA)
Information Science Reference • ©2019 • 356pp • H/C (ISBN: 9781522583226) • US $185.00

Handbook of Research on Emerging Practices and Methods for K-12 Online and Blended ...
Tina Lane Heafner (University of North Carolina at Charlotte, USA) Richard Hartshorne (University of Central Florida, USA) and Richard Thripp (University of Central Florida, USA)
Information Science Reference • ©2019 • 505pp • H/C (ISBN: 9781522580096) • US $245.00

Cases on Service Delivery in Special Education Programs
Pam L. Epler (Grand Canyon University, USA)
Information Science Reference • ©2019 • 431pp • H/C (ISBN: 9781522580690) • US $185.00

Cultivating Inclusive Practices in Contemporary K-12 Education
Johnny R. O'Connor Jr. (Lamar University, USA)
Information Science Reference • ©2019 • 308pp • H/C (ISBN: 9781522557272) • US $175.00

Handbook of Research on School Violence in American K-12 Education
Gordon A. Crews (The University of Texas at Rio Grande Valley, USA)
Information Science Reference • ©2019 • 610pp • H/C (ISBN: 9781522562467) • US $275.00

Handbook of Research on Science Literacy Integration in Classroom Environments
Chih-Che Tai (East Tennessee State University, USA) Renee M. R. Moran (East Tennessee State University, USA) Laura Robertson (East Tennessee State University, USA) Karin Keith (East Tennessee State University, USA) and Huili Hong (Towson University, USA)
Information Science Reference • ©2019 • 383pp • H/C (ISBN: 9781522563648) • US $235.00

Engaging STEM Students From Rural Areas Emerging Research and Opportunities
Reenay R.H. Rogers (University of West Alabama, USA) and Yan Sun (Mississippi State University, USA)
Information Science Reference • ©2019 • 205pp • H/C (ISBN: 9781522563419) • US $145.00

For an entire list of titles in this series, please visit:
https://www.igi-global.com/book-series/advances-early-childhood-education/76699

IGI Global
DISSEMINATOR OF KNOWLEDGE

701 East Chocolate Avenue, Hershey, PA 17033, USA
Tel: 717-533-8845 x100 • Fax: 717-533-8661
E-Mail: cust@igi-global.com • www.igi-global.com

Editorial Advisory Board

List of Reviewers

Table of Contents

Detailed Table of Contents

Inspiring learners toward career options available in STEM fields (science, technology, engineering, and mathematics) is important not only for economic development but also for maintaining creative thinking and innovation. Limited amounts of research in STEM education have focused on the population of students enrolled in religious and parochial schools in urban settings; yet given the historic conflict between religion and science, this large sector of American education is worthy of examination. This chapter incorporates Gottfredson's Theory of Circumscription and Compromise as it relates to occupational aspirations, Bem's Gender Schema Theory to explain the role of gender in career expectations, and Crenshaw's Intersectionality Theory as it pertains to religion and urban location as group identifiers. Practical interventions for encouraging young students to consider STEM careers are discussed.

This chapter provides insights to common barriers to conducting STEM professional development in an urban setting. The chapter will illustrate the complexity of integrating changes to STEM teaching practices from a systems perspective, from the lens of a researcher-practitioner conducting an 18-month study of professional development in a K-8 school in a metropolitan, public school setting. Specific challenges along themes related to the classroom level, the school level, and the district level will be discussed. Finally, the chapter concludes with implications and suggestions for future professional development providers, and schools and district leaders who are considering implementing instructional and curricular change related to STEM education.

Foreword

This is an important book. It joins educators from diverse backgrounds and institutions who offer ways to overcome rules, regulations, policies and procedures that use gender, wealth, and race to achieve repressive outcomes. Recognizing and removing barriers, stimulating interest, offering innovative learning techniques, are all necessary to produce the talent our nation requires to maintain its global competitiveness and to ensure that all students achieve their highest levels of human potential. In the information age, STEM talent goes a long way in those pursuits.

However, America is its own worst enemy when it comes to STEM education. The policies of wealth privilege, White advantage, and Black oppression, belie any notion of opportunity based on merit. The STEM pipeline that is universally desired by industry is fraught with clogs and leakage.

More than three-quarters of the wealth in this country is controlled by just a tenth of its people – a tiny sliver of the population who are nearly all White, hold the highest paying jobs, and live in "traditional" families that are often headed by men. Their disproportionate power results from a birthright that's been handed down from generation to generation, creating a closed competition in which mediocrity is in oversupply and merit yields little reward. Ask yourself, if the boss's son and a more talented poor woman or person of color were up for the same STEM job, who would win?

Poverty in the education arena is a hazardous condition, especially if you are urban and poor, Black or Brown. Urban public school systems consign those students to limited or misdirected resources, and their educational ineffectiveness is shrouded in endless cycles of "reform." Urban youth are perpetually on the wrong side of the so-called "achievement gap." Their STEM potential is systematically depleted.

Most of our nation's human capital lies not in the top 10% who are privileged, but in the 90% who are seeking a fair opportunity to succeed. Our role as educators -- especially urban educators -- is to develop and identify that potential no matter how deeply it is hidden or suppressed. The collective efforts of the authors presented here answer that calling. K-12 STEM Education in Urban Learning Environments should be required reading for every educator in every city in America.

Ronald Mason
University of the District of Columbia, USA

Preface

INTRODUCTION

Science, Technology, Engineering, and Mathematics (STEM) education has become a growing focus for the sustainability and productivity of the United States (Capraro et al., 2016). Numerous reform efforts and initiatives to increase engagement in STEM learning, degrees, and careers exist—some with successes and others with challenges. Within urban education, a unique set of challenges exists that may at times hinder availability of resources to support STEM, access to STEM, rigorous and effective implementation of STEM, and encouragement of participation in STEM. The research literature highlights the nuances of integration of STEM in urban learning environments (see Bullock, 2017), and some of the challenges that, in many ways, go unaddressed or create additional roadblocks to success in preparing adolescents to enter the STEM workforce. What remains overwhelmingly is a disparity between equal interest in STEM learning, access to STEM resources, and equal representation of minorities and females within STEM fields in general (National Science Foundation, 2017). Thus, a resource is needed that brings together various stakeholders in an effort to work towards a solution.

This book, *K-12 STEM Education in Urban Learning Environments*, allows teachers, scholars, urban community advocates, and other concerned stakeholders the opportunity to engage in research capturing the enduring issues and concerns surrounding the current K-12 STEM urban STEM education landscape. Our goal is to deepen theoretical, conceptual, and practical insights surrounding K-12 STEM, with a specific focus on urban learning environments, through surveying the perspectives of STEM educators and researchers from diverse backgrounds and experiences.

This book's emphasis on urban learning environments allows for a critical look at (1) current barriers for STEM implementation in urban K-12 schools, (2) race and gender dynamics in STEM education, (3) existing policies surrounding STEM education, (4) challenges in STEM teacher preparation, and (5) policies and practices that attend to and/or disrupt equitable, inclusive, and readily available resources for STEM education. Contributing authors for this volume hail from both professional and community spaces to share their passion for strengthening the impact of STEM education in urban school contexts.

THE CHALLENGES WITHIN URBAN EDUCATION

The term "urban education" is frequently used as a euphemism to describe the educational contexts of "Black, Brown and poor children" (Howard & Milner, 2014, p. 201). In addition to this, urban education is often stigmatized by a lack of internal and external resources for students, teachers, and communities. Along with high administrative and teacher turnover, low test scores, and neighborhoods devastated by crime. As such, urban schools are often last in terms of opportunities and resources related to STEM education, leaving many urban teachers ill equipped to provide urban students with the STEM learning experiences they deserve (Bullock, 2017). Efforts have been made at the university level in teacher preparation programs to address the disadvantages pre-service teachers may face in terms of teaching STEM in urban contexts. Although these university-level efforts have been successful in many ways (see Apugo, Wendt, Deksissa, & Peters, 2018), there are still gaps that go unaddressed: mentorship, instructional resources, and funding. Urban education boasts a large population of diverse learners, yet fragmented efforts to ensure equitable STEM education for learners continues to fall short for urban students and communities (Bullock 2017).

THE CURRENT LANDSCAPE OF K-12 STEM EDUCATION

Rigorous K-12 STEM education is often regarded as one of the key levers towards ensuring the United States' visibility as a globally competitive nation (Icel, 2018). K-12 school systems across the country have made substantial efforts to focus on STEM intensive curriculum initiatives in hopes to broaden student participation in STEM related fields. For many urban schools, the intentional integration of STEM has not been an easy task. Many K-12 schools occupying urban city spaces are challenged with resource deficits and access to STEM geared after-school programs to supplement students' interests in STEM. Along with these challenges, issues of racial and gender bias have crept into the forefront of the conversation when discussing who "belongs" in STEM (Bullock, 2017). Minimal efforts have been made to disrupt the existing STEM education narrative in K-12, although stakeholders agree that access to high quality STEM education in K-12 is a crucial asset to students' future financial success (Sondergeld, Walten, & Johnson, 2016).

The questions remain—What are we doing to support STEM education in urban K-12 learning environments? What is working? What can be improved? Are underrepresented and marginalized populations supported adequately in STEM? This book begins to address these questions and, as is our hope, can encourage open and honest discussion regarding what we need to and can do better.

CONTENT OVERVIEW

This book is presented in three sections. Section 1, "Understanding the Current Landscape of Urban K-12 STEM Education," includes two chapters. In Chapter 1, the authors discuss the issues that often plague urban schools and reform efforts through STEM. The authors propose that underserved urban schools are a key platform towards the diversification of STEM. In Chapter 2, the author incorporates Gottfredson's (1981) Theory of Circumscription and Compromise, Bem's (1981) Gender Schema Theory, and Crenshaw's (1989) Intersectionality Theory to discuss how themes of gender and religion can be used to create interventions and bolster students' interest in STEM careers.

Section 2, "Urban K-12 STEM Teacher Preparation and Professional Development," provides new directions for teacher preparation and professional development endeavors. Chapter 3 offers professional development considerations from an ecological systems perspective, while Chapter 4 offers ways to incorporate the power of analogy into instructional coaching among secondary teachers. In Chapter 5, the authors explain how the creation and implementation of language and literacy modules impact student outcomes, including meaningful participation, in K-12 science classrooms.

The three chapters in Section 3, "Strategies and Practices for Improving Urban K-12 STEM Education," closely interrogate issues related to gender, identity, equity, inclusion, and racism. In Chapter 6, authors make the case for strategically increasing STEM interest among minority girls. In Chapter 7, the need for diversified STEM education and the alignment of secondary education mathematics teacher preparation courses with urban K-12 mathematics initiatives is discussed. Finally, in Chapter 8, the authors engage readers in the impact of micro-aggressive behavior regarding underrepresented students in STEM and strategies for both recognizing and reducing divisive and detrimental behaviors.

A brief description of each of the chapters follows:

Chapter 1: Urban STEM Education – A Vehicle for Broadening Participation in STEM

The value in understanding that urban students are crucial components towards broadening participation in STEM is urgent. Authors suggest that stakeholders rise up to overcome the burdensome challenges present in urban education in order to create opportunities for long-term change in STEM education and STEM careers. Underserved students often lack STEM based educational opportunities, which tends to impact underserved student communities the most. Associated barriers and challenges are discussed in order to capture the scope of commitment needed from stakeholders. The chapter names five pressing barriers to equitable STEM education, along with potential solutions to addressing the barriers presented.

Chapter 2: STEM Career Interest at the Intersection of Attitude, Gender, and Religion

Using Gottfredson's (1981) Theory of Circumscription and Compromise, Bem's (1981) Gender Schema Theory, and Crenshaw's (1989) Intersectionality Theory the chapter considers the role of religion and gender in encouraging students to consider STEM careers. The chapter introduces a historical background of STEM, career interests and attitude, and intersecting attitudes. A thorough review of (1) attitudes in STEM, (2) interest in STEM, (3) self-efficacy and student motivation, (4) STEM career identity, (5) gender and STEM interest, and (f) STEM education in urban Christian schools. The author discusses the value of students' awareness using unique interventions.

Chapter 3: Challenges to Implementing STEM Professional Development From an Ecological Systems Perspective

Relevant, practical, and research based professional development opportunities are essential elements for great teaching. However, professional development opportunities that focus on teachers' growth and development in STEM can be a challenging find for many educators, especially those teaching in urban school communities. The author in "Challenges to Implementing STEM Professional Development from an Ecological Systems Perspective" discusses the hurdles present for comprehensive STEM teaching practices through a systems framework. This chapter considers the often complex nature of urban school environments as a critically important aspect of understanding the need for a multi-dimensional approach to resource advocacy, accessibility, and teacher preparation.

Chapter 4: The Use of Instructional Coaching and Analogy to Enhance STREAM Professional Development for Teacher Quality Improvement

Targeted professional development endeavors as it relates to STREAM (STEM with integrated Reading and Arts) may facilitate the continuous growth of teachers and students in urban school contexts. Authors in Chapter Four introduce a professional development consortium that supports STREAM focused professional development as a United States Department of Education (USDOE) Teacher Quality Improvement grant funded program. The chapter makes the call for innovative instructional coaching for improved teacher effectiveness.

Chapter 5: Bolstering Content by Promoting Language and Literacy in the Urban Science Classroom

Authors in Chapter 5 present a National Science Foundation (NSF) funded study that examines middle school STEM teachers' preparedness to work in urban school districts--particularly schools with a high proportion of linguistically and culturally diverse students. The chapter engages research around (1) STEM teacher preparation, (2) underrepresented students and STEM access and/or opportunity, (3) broadening participation in STEM, (4) urban teacher expectations for STEM education and/or classrooms, (5) racial/ethnic barriers to STEM education, (6) inclusivity and STEM education, and (7) anti-bias teaching. In addition, authors discuss how components of the NSF study enhanced urban teachers' knowledge of linguistically and culturally diverse students.

Chapter 6: Gendered and Racial Microaggressions in STEM – Definitions, Consequences, and Strategies Urban Elementary School Professionals Can Use to Combat Them

Microaggressions act as barriers to STEM education for underrepresented students. The impact of microaggressions related to STEM and "who belongs" can significantly influence student interest in STEM. Microaggressions related to gender and ethnicity are a constant hurdle for girls and other students of color. Chapter authors make the case that the weight of micro-aggressive behavior may be felt even more significantly among students attending urban schools. The consequences of both macro- and microaggressions on student learning and advances in STEM are discussed, along with recommendations for addressing micro-aggressive behavior.

Chapter 7: Secondary Education Mathematics and UDC Expectations – Do They Align?

Students' success in college mathematics depends heavily on the mastery of Common Core State Standards mathematics (CCSSM) standards in K-12. This particular study looks at the transitional factors between secondary and post-secondary math education as it relates to the CCSSM. In this chapter, the authors argue that K-12 teachers in urban areas may have fewer resources to support the math content mastery of students with diverse learning needs and abilities; thus, leaving urban students unprepared for college level mathematics. Authors ask the question "what's next for urban teachers and universities"?

Chapter 8: Improving STEM Career Aspirations in Underrepresented Populations – Strategies for Urban Elementary School Professionals

Authors in Chapter 8 highlight that the steady increase for STEM jobs will call for a more diverse group of STEM professionals—which must include women and minorities. The chapter discusses the impact of underrepresented women and minorities in STEM related careers, along with possible solutions to this ongoing crisis. Authors suggest that K-12 schooling environments work to expose minority girls to positive STEM role models in order to foster a sense of belonging, interest, and aspiration. Gottfredson's (1981) theoretical framework is used to explain and identify barriers for minority girls in schools. Moreover, a list of evidence-based strategies geared towards facilitating positive engagement and self-perception for minority girls is provided.

AUDIENCE

K-12 STEM Education in Urban Learning Environments speaks to all STEM stakeholders occupying either K-12 or university settings. This includes: scholars, administrators, policy-makers, STEM education instructors, researchers, students, and parents. Given the variety of backgrounds of the chapter authors, as well as their diverse perspectives, this book provides insight regarding where we are in implementing STEM within the urban K-12 environment, what challenges remain, and where future initiatives and policy must take us in order for STEM opportunities to be accessible for all students.

Jillian L. Wendt
University of the District of Columbia, USA

Danielle L. Apugo
University of the District of Columbia, USA

REFERENCES

Apugo, D., Wendt, J. L., Deksissa, T., & Peters, M. (2018). The science behind support: An informed perspective on preparing urban science teachers for diverse 21st century classrooms. In A. S. Burtin, J. S. Fleming, & P. Hampton-Garland (Eds.), Changing urban landscapes through public higher education (pp. 97-115). Academic Press.

Bem, S. L. (1981). Gender schema theory: A cognitive account of sex typing. *Psychological Review*, *88*(4), 354–364. doi:10.1037/0033-295X.88.4.354

Bullock, E. C. (2017). Only STEM can save us? Examining race, place, and STEM education as property. *Educational Studies*, *53*(6), 628–641. doi:10.1080/00131946.2017.1369082

Capraro, R. M., Capraro, M. M., Scheurich, J. J., Jones, M., Morgan, J., Huggins, K. S., ... Han, S. (2016). Impact of sustained professional development in STEM on outcome measures in a diverse urban district. *The Journal of Educational Research*, *109*(2), 181–196. doi:10.1080/00220671.2014.936997

Crenshaw, K. (1989). Demarginalizing the intersection of race and sex: A black feminist critique of antidiscrimination doctrine, feminist theory, and antiracist politics. *University of Chicago Legal Forum*, *1989*, 139–167.

Gottfredson, L. S. (1981). Circumscription and compromise: A developmental theory of occupational aspirations. *Journal of Counseling Psychology*, *28*(6), 545–579. doi:10.1037/0022-0167.28.6.545

Howard, T. C., & Milner, H. R. (2014). Teacher preparation for urban schools. In R. Milner & K. Lomotey (Eds.), *Handbook of urban education* (pp. 199–216). New York, NY: Routledge.

Icel, M. (2018). Implementation of STEM policy: A case study of a STEM-focused urban charter school. *Journal of STEM Education*, *19*(3), 7–13.

National Science Foundation. (2017). *Women, minorities, and persons with disabilities in science and engineering*. Retrieved from https://www.nsf.gov/statistics/2017/nsf17310/

Sondergeld, T. A., Johnson, C. C., & Walten, J. B. (2016). Assessing the Impact of a Statewide STEM Investment on K-12, Higher Education, and Business/Community STEM Awareness Over Time. *School Science and Mathematics*, *116*(2), 104–110. doi:10.1111sm.12155

Acknowledgment

We would like to first and foremost thank our contributing authors for their dynamic contributions. We deeply appreciate their dedication to the process of making this publication a reality. Next, we wish to acknowledge the valuable contributions of the reviewers regarding the improvement of quality, coherence, and content presentation of chapters. Many of the authors also served as referees; we highly appreciate their commitment to this project. We also want to honor our editorial advisory board (EAB) for their commitment, patience, and dedication to reviewing submissions with timely and valuable feedback. Their diligence in ensuring a rigorous and valuable project is very appreciated. Finally, we wish to thank our friends, family, and colleagues, near and far, who have supported us through the process of compiling and editing this publication. Without your encouragement and support, this project would not have been possible!

Jillian L. Wendt
University of the District of Columbia, USA

Danielle L. Apugo
University of the District of Columbia, USA

Section 1

Understanding the Current Landscape of Urban K–12 STEM Education

Chapter 1
Urban STEM Education:
A Vehicle for Broadening Participation in STEM

Brandy Huderson
University of the District of Columbia, USA

Ashley Huderson
American Society of Mechanical Engineers, USA

ABSTRACT

Students in underserved urban communities often lack access to educational opportunities in STEM (science, technology, engineering, and math). Evidence reveals a lack of STEM access in urban environments, displaying a clear pattern impacting predominantly impoverished neighborhoods the hardest. The issue of STEM education in urban schools reflects an increased interest in expanding and diversifying the future workforce to compete in a STEM-centered future. STEM has been positioned as a critical part of urban education reform efforts and a potential vehicle to broadening participation in STEM. In various US cities, schools labeled as failing are being repurposed as selective STEM-intensive academies to build a STEM education infrastructure. In this chapter, the authors discuss the barriers associated with urban STEM education and how this underserved population is a vital vehicle to increase diversity in STEM.

DOI: 10.4018/978-1-5225-7814-7.ch001

INTRODUCTION

The ability of this nation to support a growing economy, maintain health and human services, and ensure national security depends upon a vibrant, creative, and diverse science, technology, engineering, and mathematics (STEM) workforce. The end of World War I left much of non-Russian Europe financially disseminated, which set in motion a series of steps that would eventually lead to the United States (U.S.) becoming a global economic powerhouse (Focus Economics, 2018; Tooze, 2008). This position has been defined by a highly skilled and competent STEM workforce, leading in innovation and competition (Xie & Shauman, 2015; Focus Economics, 2018). However, in recent years developing trends have resulted in dramatic shifts in the U.S. STEM workforce resulting in the U.S. now ranking 29th in the developed world for producing STEM-related college degrees according to the 2015 Organisation for Economic Co-operation and Development (OECD) Science, Technology, and Industry Scoreboard (Organisation for Economic Co-operation and Development, 2018). Despite decreases in the number of STEM-related degrees conferred in the U.S., there remains a steady demand for skilled, American STEM workers in response to the ever-changing landscape of technology fueling innovations in how goods and services will be manufactured, engineered, and delivered (Carnevale, Smith, & Melton, 2011). Demand is projected to be greatest for engineers (mechanical and electrical), computer programmers and software developers, and end user support (Carnevale, Smith, & Melton, 2011; Education Week, 2018). Presently, there is growing concern about the nation's ability to sufficiently meet these workforce demands. It has been stated that in large metropolitan areas ~30% of all employment needs are STEM-based while only a fraction of the population have the needed STEM training or education (Rothwell, 2013). According to the National Science Board's 2016 Science & Engineering Indicators report, the U.S. is faced with the following workforce challenges in STEM (National Science Board, 2016):

- An underrepresentation of women, African Americans, and Hispanics in STEM jobs at rates lower than their U.S. population representation;
- An aging STEM workforce (median age of STEM workers has risen from 41 to 43 years old between 1993 and 2013);
- A STEM workforce disproportionately foreign-born compared to other occupations;
- And a demand for science and engineering workers

These identified challenges can be directly articulated to the dearth in effective implementation policies and best practices that are designed to ensure and improve equal access to STEM education. As a result, invested stakeholders struggle to provide students with the STEM experiences required to be competitive for the 21st century, regardless of college and career aspirations (Indiogine, 2015). In particular, state and local education agencies and school-level educators struggle to close persistent achievement gaps in core subjects, specifically in mathematics and science (Indiogine, 2015; Morgan, Morgan, Farkas, Hillemeier, & Maczuga, 2016). While STEM faces many nuanced challenges, by far the biggest and most agreed upon, is lack of a sufficient STEM-skilled talent pool (Carnevale, Smith, & Melton, 2011). This is particularly vexing considering that the number of STEM-related jobs has increased by ~30% in comparison to non-STEM jobs over the last decade. Past estimations predicted that more than two million STEM jobs would go unfilled. (Carnevale, Smith, & Melton, 2011; Rothwell, 2014).

Over the ten year period from 2001 to 2011, there was no more than a five percent increase in the number of science and engineering degrees awarded (bachelor's, master's, and doctoral) at the individual state level (National Science Board, 2014). Furthermore, over this same 10 year period, the proportion of total degrees awarded in science and engineering remained static (29.6% and 29.3% in 2001 and 2011, respectively) (National Science Board, 2014). This is concerning, because as previously stated, during this same time frame STEM-related jobs increased by three times. Furthermore, in 2014 the United States Census Bureau reported that ~74% of individuals having received a bachelor's degree in a STEM-related field were not employed in a STEM field (U.S. Census Bureau, 2014).

As of 2016, there was an increase in the number of master's and doctoral degrees in STEM being awarded as well as the number of STEM jobs filled (National Science Board, 2016). However, further examination of this data shows that this increase is due in part to a significant rise in the number of international students receiving STEM-related degrees (Carnevale, Smith, & Melton, 2011; National Science Board, 2016). While it is tempting to place the blame of a low STEM-trained workforce on lack of retention of STEM-focused students in the undergraduate process, data published by the National Science Foundation showed ~60% of students entering four-year universities in STEM programs remained in the major, receiving a bachelor's degree (National Science Board, 2014). Attempting to understand the data and trends explaining the state of STEM in the United States can be a bit confusing, but in the end, at least one conclusion seems clear:

The number of students entering post-secondary schools committed to STEM-related majors needs to increase in order to provide a viable domestic STEM workforce (Carnevale, Smith, & Melton, 2011; U.S. News/Raytheon STEM Index, 2016).

Preparing students for post-secondary education is generally seen as the role and job of K-12 education. However, assessment of the nation's schools using Education Week's Chance-for-Success Index, analysis of school finances, and K-12 Achievement Index resulted in an overall grade of "C" (74.2%) for 2017, which was little changed from 2016 (74.4%) (Education Week Research Center, 2016; Lloyd & Harwin, 2016). Urban schools faired much worse. In its ground breaking report published in 2015, the Center on Reinventing Public Education (CRPE) at the University of Washington reported minimal proficiency gains in math and reading (DeArmond, et al., 2015). In the 18 cities examined, schools in the lower 5th percent struggled to improve their standings, with most failing to improve over the course of three years (DeArmond, et al., 2015). With an estimated 30% of all U.S. school-aged children participating in urban education systems, failing and lagging schools systems have far reaching implications for increasing the U.S.-born STEM workforce (Urban Education in America, 2013). This chapter aims to provide a detailed explanation of urban STEM education from both a historical and contemporary context. We will examine challenges urban STEM education faces as well as the systemic and institutional barriers urban students face. We will also provide examples of STEM interventions that have had success in improving urban students' proficiency, interest, and persistence in STEM-related educational tracks.

BACKGROUND

In order to effectively address the challenges and barriers faced in urban STEM education we must first clearly articulate what urban education is. The importance of this was no clearer than in Milner's (2012) account of a visit to the Midwest. In his account, Milner states that educators considered their service to be "urban" based on the student population and the socio-economic status (Milner, 2012). Over the years, scholars in the field have sought to provide an expanded and more inclusive description of urban education. In general, we can discuss public urban education as learning centers that exist in metropolitan areas, where there is a tendency for increased cultural and linguistic diversity as well as socio-economic diversity; these learning centers tend to have large enrollment, and larger student-to-teacher ratios (Watson, 2011; Urban Education in America, 2013). As with any highly generalized description or grouping, the local landscape of public urban education is highly interdependent on the microenvironment of the school. For example, there may exist schools where there is a more homogenous student population with variety in the socio-economic status of the students, or the student body population is highly diverse with their commonality being poverty status (Milner, 2012).

In 2013, the National Center for Education Statistics' report *Urban Education in America* estimated that ~80% of US students, roughly 50 million students, ranging from Pre-K to 12th grade, were in public schools (Urban Education in America, 2013). Of that 80%, ~30%, nearly 15 million students, were located in city or urban locales. Further examination of city and urban public schools reveals that only 13%

of these schools are deemed to be "low-poverty" schools, meaning that less than 25% of the student body is eligible for free or reduced-price lunch (FRPL) (The Condition of Education: Concentration of Public School Students Eligible for Free or Reduced-Price Lunch, 2018). Over 60% of city and urban public schools have at least 50% of the student body eligible for FRPL, adding to the complex context of poverty to the majority of American urban public schools (The Condition of Education: Concentration of Public School Students Eligible for Free or Reduced-Price Lunch, 2018).

Students in these areas are more often than not at the receiving end of harmful pre-conceived ideas about both their desire to learn as well as their ability to learn. "Urban students come from an environment where they can't see the value of education" (Watson, 2011). Often times, it is these negative associations with urban schools and their students that provides the first and, at times, the most formidable barrier to providing effective STEM education in urban settings. Across the nation, urban schools have historically struggled to hire and retain teachers in comparison to suburban and rural areas (Lippman, Burns, & McArthur, 1996). More than twenty years ago, published reports stated the negative impact that chronic problems such as low teacher morale and high teacher absenteeism, had on student progress (Lippman, Burns, & McArthur, 1996). A review of published data found that many preservice teachers, who did not come from urban areas had limited knowledge and experience with cultures and languages apart from their own (Hill, Friedland, & Phelps, 2012). Teacher candidates perceived and expected urban students to universally exhibit behavioral problems, to be intellectually limited, and that the actual schools would suffer from chronic under-funding and inadequate resources (Hill, Friedland, & Phelps, 2012). These same teacher candidates were surveyed following immersive field experiences that placed them in direct contact with urban students and the urban educational system at which point many of their pre-conceived notions were proven false (Hill, Friedland, & Phelps, 2012).

URBAN STEM EDUCATION

Over the years, initiatives to promote and prioritize STEM education have resulted in few achievement gains in urban areas. In a 2012 report, the U.S. Government Accountability Office (GAO) reviewed STEM education programs across 13 federal agencies and reported an 83% overlap between program services and target populations (United States Government Accountability Office, 2012). Additional reviews of federal STEM programs have cited a general lack of cohesiveness of what urban STEM education should be at both the national and local levels and across all educational levels (President's Council of Advisors on Science and Technology,

2010). This has been further compounded by a lack of clear goals for urban STEM education programs and no effective assessment of implemented programs (President's Council of Advisors on Science and Technology, 2010) (United States Government Accountability Office, 2012; Bullock, 2017) (United States Government Accountability Office, 2012). The generalized chaos around STEM education has been shown to affect poor, under resourced urban and rural areas the hardest where STEM education is seen as a key to better paying jobs and a relief for economically depressed areas (Bullock, 2017). According to Eisenhart and colleagues (2015):

A key aspect of STEM reform at the national level is to improve education and life outcomes for students without these privileges. Especially in cities with large low-income and minority populations, STEM-focused schools are being promoted as a means to distribute opportunities more broadly to historically disenfranchised groups...increasing the quality of STEM education, (National Research Council, 2011) a national priority. (p. 764)

A cornerstone in urban STEM education is the creation of STEM-focused schools. These educational units are a proven way to increase proficiency in science and mathematics as well end the cycle of low-paying jobs in economically depressed urban areas (Eisenhart, et al., 2015; Bullock, 2017). Recently, there has been a significant increase in the number of STEM-focused schools around the nation. This is no doubt due in part to the President's Council of Advisors on Science and Technology (PCAST) call for the establishment of no less than 1000 STEM-focused schools by 2020 (President's Council of Advisors on Science and Technology, 2010). This increase has also been supported by another Obama-era action item, the Race to the Top initiative, which provided federal grant dollars for innovation education reform (U.S. Department of Education, 2009).

Problems With Urban STEM Education

Urban STEM education and the associated ongoing reforms are tightly linked to the overall urban education reforms designed to improve the quality of education in urban areas through the creation of strategic partnerships with institutions of higher education, industry stakeholders, and other formal and informal entities (U.S. Department of Education, 2016). As far back as the National Defense Education Act of 1958, there has been a national agenda for increased proficiency in science and math (Carmichael, 2017). Framed as an issue of economics and global competition, science and math education has been seen as the key to the U.S. remaining a global leader in technology and a way to ensure its workforce. Despite this constant push, U.S. students continue to lag behind our international counterparts. According to the most recent results from the Program for International Student Assessment (PISA), U.S. students are average in science but are below average in math in comparison to their international counterparts-- 24th and 38th, respectively, of 71 ranked countries (http://www.oecd.org/pisa/, 2016). In 1991, the National Science Foundation (NSF) proposed a large-scale, top-down model for improving science and math performance nationally. Implementing a Statewide Systemic Initiative (SSI), the NSF granted competitive states up to $10 million dollars to institute statewide programming focused on science and math (Zucker, Shields, & Adelman, 1998). Although not specifically focused on urban areas, SSI structure required policy and programmatic design and implementation at the district level and stressed the importance of clear and coherent objectives at both the regional- and state level. Despite previously described initiatives and increased rigor in STEM focused programs, urban students continue to lag behind the national average in both science and mathematics (Camera, 2015; Cavanagh, 2006). A recently published evaluation of urban STEM education reform in two major cities provided interesting insight into the effectiveness of targeted reform efforts. Weis et. al (2015) reported that even with the creation of "inclusive STEM-focused" schools, students are still plagued by systemic inequality and failure to provide adequate infrastructure for program success.

One example of an urban school district would be that found in our nation's capital. In 2015, the Washington, DC public education system (DC Public Schools and Public Charter Schools) reported, based on standardized test scores, that only 30% of the third grade students were "meeting expectations" or greater in math; this percentage steadily declined from fourth to seventh grade, with a small rebound for eighth graders (2015 DC PAARC Results, 2015). When data was expressed as a function of racial or ethnic groups or students from special populations, significant disparities were evident. As of 2016, across all ethnic groups, only 24% of DC's public education students were college-ready in math (Office of the State Superintendent of Education, 2016). This number increased to ~27% in 2017 (Office of State Superintendent of Education, 2017). Approximately 70% of white students were college ready in math in comparison to 17% of black or African American students and 20% Hispanic students (Office of the State Superintendent of Education, 2016). Looking at the data independent of race still provides a problematic snapshot of DC's public educational system. In 2016, only 17% of DC's students classified as *Economically Disadvantaged* were college "on track or ready for college" in math; similar results were reported for *English Language Learners* (ELL; 18.5%) while only 13% of *At-Risk* students were on target (Office of the State Superintendent of Education, 2016). There were modest improvements the following year with the number of *Economically Disadvantaged* math ready increased to 19%, and similar trends seen for *ELL* (21%) and *At-Risk* (14%) students (Office of State Superintendent of Education, 2017). Separation of PARCC data between the two sub fractions of the

DC Public Educations system, showed that while *Economically Disadvantaged* and *At-Risk* students performed better in DC Charter schools in 2016 and 2017(see Table 1), there was little improvement in college readiness (Office of State Superintendent of Education, 2017). The trend of declining proficiency from third to eighth grade was not changed, however, leaving the system ranked near the bottom (49th) in national rankings (McAnn, 2018). DC's public education system has been fighting an uphill battle but there have been noticeable improvements in various areas of the education continuum. Although the overall grade did not improve (C-), DC has seen improvements in "chances for success" (C+ to B-) and K-12 achievement (F to D) over a four year span from 2013 to 2017 (Lloyd & Harwin, 2016).

BARRIERS SURROUNDING URBAN STEM EDUCATION

Understanding regional trends in public education is vital for impactful implementation of interventions. It is also necessary to consider nationally identified barriers for students in STEM. In particular, the role that these barriers play in maintaining or even widening the achievement gap between students of color and their white counterparts is important to examine. In this text, we will focus on structural, cultural and psychological barriers within the K-12 educational system. Based on a review of the literature, the following areas present structural barriers to urban STEM education: Absence of supplies and resources, minimal investment in teacher's professional development, inadequate preparation of students, deficiency in resources from the school system, and insufficient content delivery and assessment.

Table 1. DC PARCC Data, 2016 and 2017. Percentage of students "on track or ready for college. Data sourced from 2017 PARCC Results, 2017. (Office of State Superintendent of Education, 2017)

	DC Public Schools		DC Public Charter Schools (DCPC)		DC Public Schools (DCPS)	
	2016	2017	2016	2017	2016	2017
Economically Disadvantage Students	17%	19%	22%	22%	14%	17%
At-Risk Students	13%	14%	18%	17%	10%	12%
English Language Learners	19%	21%	15%	15%	20%	23%

Absence of Supplies and Resources

Lack of supplies and resources in urban schools has been linked to diminished student engagement and achievement (Banerjee & Lamb, 2016). Highly skilled and trained teachers in high poverty urban schools have generally reported having to work with outdated supplies and curriculum in over packed classrooms. The lack of supplies includes outdated textbooks, computers, and technology; and inadequate or nonexistent science equipment, materials, and labs (Hudley, 2013). According to U.S. News and World Report, millions of students across the country live in what we call STEM deserts – school communities without access to rigorous and engaging math and science courses (Matthew, 2017). In addition, these schools offer less college-preparatory and advanced placement courses in comparison to schools serving more advanced populations (Freel, 1998). Physical conditions in urban schools have long been used as a predictor of academic engagement and performance (Lewis, et al., 1999). However, basic materials — including textbooks, science equipment, and desks — were generally in poor condition or absent all together. Overall, physical and academic conditions in most urban schools often leaves them labeled as places of developmental risk as oppose to institutions of learning to enhance student development (Hudley, 2013).

Minimal Investment in Teacher's Professional Development

Professional development of teachers is an essential component linked to improving teaching and learning (Tirozzi & Uro, 1997). It has well been established that a direct correlation exists between education of students and knowledge and influence of teachers (National Commission on Teahing and America's Future, 1996; Banerjee & Lamb, 2016). Research studies have identified effective professional learning as a critical component of school success (Wei, Darling-Hammond, Andree, Richardson, & Orphanos, 2009). However, despite these studies, there has been a national lack of concern to ensure educators are provided effective professional learning, thus the structure for teacher professional development remains fragmented (Hill, 2009). This fragmented system is even more evident in urban schools as it relates to STEM teachers. Continual professional development activities in mathematics and science are critical to improving the content knowledge and skills of elementary teachers and mathematics and science teachers at the middle and high school levels (Darling-Hammond, Hyler, & Gardner, 2017; Hibpshman, 2004). According to (Mervis,

2011) "anything that dilutes those ingredients—budget cuts, poor teacher preparation and professional development, a disregard for low-achieving students, to name three factors—will lower the chances of success." There is a critical need for new educators to be mentored by expert mentor educators, to ensure they learn to teach effectively. Many school districts assign new teachers into the toughest classes with no assistance during their first critical months of teaching (DePaul, 2000). However, when a school district includes mentoring by an experienced teacher, opportunities to collaborate with colleagues and gets assistance in managing assignments, this will allow them to learn to teach effectively (DePaul, 2000).

Inadequate Preparation of Students

According to a survey by Microsoft and Harris, most college students studying for degrees in science, technology, engineering, or math make the decision to do so in high school or before (Microsoft Corp. and Harris Interactive, 2011). Only 20% say they feel that their education before college prepared them "extremely well" for those fields (Microsoft Corp. and Harris Interactive, 2011). The survey was administered to both college students pursing STEM degrees and the parents of K-12 students about their attitudes toward STEM education. The results also found that male and female students enter the fields for different reasons: females are more likely to want to make a difference, while males are more likely to say they've always enjoyed games, toys, or clubs focused on the hard sciences. According to the 2010 PCAST report, too few U.S. students are proficient in STEM (PCAST, 2010) and too few of those who are proficient pursue STEM fields. For example, of all ninth graders in the United States in 2001, only about 4% were predicted to earn college degrees in STEM fields by 2011. This loss of potential STEM talent begins well before high school. For example, in both mathematics and science, 70% of eighth graders who lack proficiency face a mounting barrier as they experience increased difficulty in STEM subjects due to lack of a solid foundation in basic skills such as algebra (Ejiwale, 2003). Among minority students who are proficient in STEM in eighth grade, 60% decide during high school that they are not interested in those subjects and only about 40% actually enter STEM majors in colleges (PCAST, 2010). According to the report, two recommendations will suffice to address these challenges: preparing students to have a strong foundation in STEM regardless of their career path and inspiring students to be motivated to learn STEM subjects so that many of them will be excited to enter STEM fields.

Deficiency in Resources From the School System

Structural barriers in K-12 STEM education are the inevitable consequence of social and economic racism which prevent students from accessing basic educational resources (Scott & Martin, 2014). One major policy based barrier is lack of access to funding. (DC State Board of Education, 2018). This is important because in these areas parents are less likely to be able to invest monies directly in the school or provide supplemental math and science instruction. Additionally, schools serving high-poverty areas are more likely to be understaffed and under resourced; these schools may not have current textbooks or technologies which negatively impacts student achievement (García & Weiss, 2017). Schools in these areas are less likely to recruit highly skilled teachers and as such have a higher percentage of unlicensed teachers or teachers providing instruction outside of their educational qualifications (Dee & Goldhaber, 2017). In addition, minority-dense schools and schools in high poverty areas were more likely to have teachers rated as "Ineffective," by their students (Gist, Bianco, & Lynn, 2018).

Insufficient Content Delivery and Assessment

The method of teaching determines the amount of knowledge that learners acquires (Onuja, 1987). The STEM educator as a facilitator should not only be knowledgeable in the subject but should also possess the basic and necessary skills with which to impart the knowledge of the subject to learners at all levels of learning (Nwanekezi & Nzokurum, 2010). A noticeable characteristic of many urban schools is that they generally have relatively few well-qualified teachers. These teachers tend to have temporary or emergency certification, teach in fields for which they lack strong subject-matter preparation ("out-of-field"), or are in their first year or two of their teaching careers (Darling-Hammond, 2000; Banerjee & Lamb, 2016). These schools also have a difficult time attracting teachers with strong qualifications, especially in core subject areas, and often are unable to keep teachers for more than a few years (Darling-Hammond, 2000). Many elementary teachers not only lack confidence to teach STEM, but also experience anxiety related to STEM education (Harper & Daane, 1998; Hansen & Gonzalez, 2014). When teaching is not effective, the learners grasp very little. Due to this position of impact, STEM educators should attempt to understand the available methods and teaching strategies available around delivering STEM content to diverse audiences. A "cookie-cutter" approach will not suffice when the mold cast does not represent all students in the continuum and only leads to disappointment and continued marginalization of underrepresented students in urban schools. The continued usage of poor pedagogies in urban K-12 classrooms, despite large pressures for teachers to change their practice, points towards systematic as well as conceptual barriers to change (Hird, 2013).

POTENTIAL SOLUTIONS TO BARRIERS

It is critical that academic interventions for teaching STEM in urban schools become a national priority. Implementation and evaluation of interventions must be sensitive to generational status and differences among students in urban schools. Below we have proposed 5 ways to potentially address the barriers associated with urban STEM education. These potential solutions include: After school programs, curriculum challenges, the culture of STEM schools, and bridge programs.

After School Programs

Evidence shows that students spend 86.7% of their time outside of the classroom (Gerber, Cavallo, & Edmund, 2001; Meguid & Collins, 2017), thus utilization of this time to increase STEM interest and engagement is beneficial. Out of school programs have increasingly been recognized as crucial components of the larger learning ecosystem for STEM. In the out of school space, participants become interested and engaged in STEM, and also develop tangible STEM skills and proficiencies, come to value these fields and their contributions to society, and begin to see themselves as potential contributors to the STEM enterprise (Krishamurthi, 2014). Given the state of STEM education and the role out-of-school programs play in the learning ecosystem, focusing on STEM out-of-school programs has the opportunity to impact and shape the interest of students traditionally underrepresented in STEM.

Some out-of-school learning programs around the nation have embraced STEM programming and are engaging children and youth in STEM learning—including those who may not otherwise be selected to, or choose to, participate in STEM programs. A recent study showed that 7 million children are participating in afterschool STEM programs in the United States (After School Alliance, 2011; After Allinace, 2014). There has been a significant effort to define frameworks or youth outcomes in afterschool STEM programs and to document evidence of impacts (The After School Alliance, 2013). These programs host varied modes of intervention, allowing educators to match learning experiences to student interests and to use project-based learning that drives home the relevance and importance of STEM in daily life. The afterschool setting gives students the opportunity to learn through solving problems and through failing—an experience crucial to research, experimentation, innovation, and developing the persistence the fields require.

Students develop an understanding of STEM concepts and skills through engagement in a wide array of learning experiences that take place both in and out of school. In developing criteria for successful informal STEM programs, the National Research Council (2015) concluded that there were three characteristics of successful informal STEM education programs:

Productive Programs Engage Young People Intellectually, Socially, and Emotionally

Productive out-of-school STEM programs provide young people with firsthand experiences with STEM phenomenon and materials, engage them in sustained STEM practices, and are aligned with participants' cultural resources and practices. In such programs, young people are engaged in firsthand, materials-rich, and place-based learning experiences that involve processes of scientific or engineering investigation and practice. Thus, productive out-of-school STEM programs engage young people in the processes of doing STEM in ways they find compelling and challenging, and develop their interest, understanding, and commitment to continue engaging in STEM learning.

Productive Programs Respond to Young People's Interests, Experiences, and Cultural Practices

Productive out-of-school STEM programs make STEM relevant to the questions that interest young people, support collaboration and leadership by young people, and train staff to support and build young people's STEM activities and interest. Productive out-of-school STEM programs are also responsive to young people's prior interests and experiences so that they can see STEM as meaningful and relevant to their own experiences and aspirations.

Productive Programs Connect STEM Learning in Out-of-School, School, Home, and Other Settings

Productive out-of-school STEM programs explicitly help young people make connections among STEM experiences in and across settings and programs, leveraging community resources and partnerships and brokering ongoing opportunities to engage in STEM learning activities. Productive out-of-school programs also help young people understand how what they experience and learn relates to learning in other settings, including school. Thus, productive out-of-school programs purposefully help young people, their parents, and others in the community capitalize on developing expertise and interests across time and setting.

In addition, there is a large body of psychological and motivational research on underrepresented students in STEM. The research on historically underrepresented students largely focuses on engaging students in school. However, recent research suggests that out of school programs are influential in engaging students in STEM and bridging out of school experiences with classroom learning (The Afterschool Alliance, 2013). The existing research on the factors that influence STEM engagement, course selection in college, and career choice suggests that early exposure to STEM activities is significantly related to those influencing factors (The After School Alliance, 2013).

Curriculum

Part of the problem arising from lack of STEM training within public schools is absence of a relevant curriculum. Although the schools have benchmarks and standards, they are often not truly relevant. Many of the education benchmarks have been adopted from a Western curriculum and not altered to be locally relevant (Altinyelken, 2015). The textbooks, if the schools even have them, are not something students can identify with (Picardo, 2016). If students cannot identify with what they are being taught, they probably will not develop an interest in that subject. Building out access to Advanced Placement (AP) STEM curriculum is a good way to expose students to the foundational content students will need to pursue four year degrees and succeed in their coursework (Schneider, Broda, Judy, & Burkander, 2014). While the value proposition for AP exams is still up for debate, especially for low-income and under-resourced districts, "passing" exams is still a significant predictor of future STEM success.

Inclusive STEM Schools

Highly selective high schools specializing in STEM education have existed for many decades, encouraging students with identified STEM talent to pursue careers as STEM leaders and innovators (Forman, et al., 2015). Over the years, these STEM schools have grown and are responsible for increasing the proportion of students who went on to study STEM in college. For example, 51% of students from STEM high-school reported majoring in science, whereas 23% of the national sample graduated with a science major (Subotnik, Tai, Rickoff, & Almarode, 2009). Despite this, STEM schools have received push-back, with opponents believing these schools take the most "talented" students out of general public schools.

Inclusive STEM schools have recently emerged as a better option. They are specialty programs (either a standalone school or "school within a school") but take a different approach than traditional STEM charter or magnet schools by accepting students not on test performance or academic record, but on an "interest" in STEM subjects (Stuessy, 2015). Inclusive STEM high schools (ISHSs) have open enrollment and are focused on preparing underrepresented students for the pursuit of advanced STEM studies. They work to provide a critical mass of nontraditional STEM students, defying stereotypes around STEM identities and work (Stuessy, 2015) .

Bridge Programs

There has been an increasing interest in increasing the retention of students in STEM and to enhance the diversity of STEM professionals (Ashley, Cooper, Cala, & Brownell, 2017; American Association for the Advancement of Science, 2015; Chen, 2013; President's Council of Advisors on Science and Technology, 2012). Due to this interest, there has been increased attention to the development of bridge programs for STEM disciplines to prepare students to meet academic demands of STEM degree programs and to address non-academic factors that may contribute to the attrition of students in STEM (Seymour & Hewitt, 1997; Ashley M., Cooper, Cala, & Brownell, 2017). Bridge programs at the precollege and college level have also demonstrated success in increasing STEM persistence (U.S. Department of Education, 2016). Bridge programs are low-cost pre-college programs that last for a few days to several weeks. They help students who may be less likely to be prepared for college coursework increase their basic subject matter understanding, increase their understanding of fundamentals and gain access to resources and community support. These programs are often intensive experiences that occur in the weeks before a student's first year of college or a transfer student's entry into a 4-year institution. Such programs are commonly referred to as summer bridge programs, boot camps, summer programs, or college prep programs.

CONCLUSION

With the prediction of more than two million STEM jobs going unfilled in the U.S. in the coming years, it is crucial that the U.S. increase its STEM-skilled workforce. For students who are or will be entering the K-12 pipeline, emphasis needs to be placed on acquiring the necessary STEM-related skills and education to successfully compete in the job market. However, roughly one-third of the U.S. K-12 student population is in urban education systems, which have historically fared much worse in student achievement and proficiency.

Attempts to remedy this situation have focused on STEM and STEM-related education programs. Driven, in part by the creation of STEM-focused schools, this was thought to be the vehicle of success for urban schools. However, lack of overall cohesiveness of both national and local programs, and rampant duplication of programs across federal agencies has led to little improvement in either urban education or urban STEM education. In addition to the previously mentioned problems, a failure to address the systemic and institutional barriers within urban STEM education has resulted in over twenty years of little to no national progress for urban K-12 students. In this chapter we evaluated five potential solutions to dealing with urban STEM education barriers: *After school programs, curriculum challenges, the culture of stem schools, and bridge programs*. All of these solutions have been previously implemented in various locales and have reported success with their target populations. It is important to note that each one of these proposed solutions focuses on a different subset of problems facing urban STEM education. This list is inclusive of K-12 educators, higher education, and both local and national communities as a whole. Understanding the barriers that specifically affect urban STEM education is paramount in (1) increasing the number of STEM-proficient students for urban areas; and (2) working to provide a sustainable, domestic STEM-competent workforce. Addressing these barriers requires a comprehensive examination of the structural, economic, and social constructs that have perpetuated the problem. Taking from past successes as well as missteps, it is plausible that urban STEM education can be reshaped with focuses on creating impactful solutions that directly address both perceived and actual barriers.

REFERENCES

Allinace, A. (2014). American After PM: Afterschool Programs in Demand. Washington, DC: Academic Press.

Altinyelken, H. K. (2015). *Evolution of Curriculum Systems to Improve Learning Outcomes and Reduce Disparities in School Achievement.* Academic Press.

American Association for the Advancement of Science. (2015). *Vision and change in undergraduate biology education: A call to action.* Retrieved from http://visionandchange.org/files/2013/11/aaas-VISchange-web1113.pdf

Ashley, M., Cooper, K. M., Cala, J. M., & Brownell, S. (2017). Building Better Bridges into STEM: A Synthesis of 25 Years of Literature on STEM Summer Bridge Programs. *CBE Life Sciences Education, 16*(4), e3. doi:10.1187/cbe.17-05-0085 PMID:29146667

Banerjee, P. A., & Lamb, S. (2016). A systematic review of factors linked to poor academic performance of disadvantaged students in science and maths in schools. *Cogent Education, 3*(1).

Bullock, E. C. (2017). *Only STEM Can Save Us? Examining Race, Place, and STEM Education as Property.* Retrieved from http://tandfonline.com/action/journalInformation?journalCode=heds20

Camera, L. (2015). *Students in Urban School Districts Lagging in Math.* Reading, MA: U.S. News.

Carmichael, C. (2017, May 13). *A State-by-State Policy Analysis of STEM Education for K-12 Public Schools*. Seton Hall University Dissertaions and Theses.

Carnevale, A. P., Smith, N., & Melton, M. (2011). *STEM*. Washington, DC: Georgetown Universtiy Center on Education and the Workforce.

Cavanagh, S. (2006). Simple Science Difficult for Urban Students to Grasp, NAEP Study Finds. *Education Week*.

Chen, X. (2013). STEM attrition: College students' paths into and out of STEM fields. Washington, DC: Statistical Analysis Report NCES 2014-001.

Microsoft Corp. & Harris Interactive. (2011). *STEM Perceptions: Student & Parent Study Parents and Students Weigh in on How to Inspire the Next Generation of Doctors, Scientists, Software Developers and Engineers.* Harris Interactive.

Darling-Hammond, L. (2000). Teacher Quality and Student Achievement. *Education Policy Analysis Archives*, *8*(1), 44.

Darling-Hammond, L., Hyler, M., & Gardner, M. (2017). *Effective Teacher Professional Development*. Palo Alto, CA: Learning Policy Institute.

DC State Board of Education. (2018). *Teacher and principal turnover in public*. Retrieved from https://dcps.aqdev.dc.gov/page/compensation-lift-and-impactplus

DeArmond, M., Denice, P., Gross, B., Hernandez, J., Jochim, A., & Lake, R. (2015). *Measuring Up: Educational Improvement and Opportunity in 50 Cities*. Center on Reinventing Public Education.

Dee, T., & Goldhaber, D. (2017). *Understanding and Addressing*. The Hamilton Project.

DePaul, A. (2000). *Survival Guide for New Teachers How New Teachers Can Work Effectively with Veteran Teachers, Parents, Principals, and Teacher Educators*. U.S. Department of Education, Office of Educational Research and Improvement.

Economics, F. (2018, November). *Economic Forecasts from the World's Leading Economists*. Retrieved from https://www.focus-economics.com/countries/united-states

Education Week. (2018, May 22). *Chart: Which STEM Jobs Are In Demand and Pay Well?* Retrieved from https://www.edweek.org/ew/section/multimedia/chart-which-stem-jobs-will-be-in.html

Education Week Research Center. (2016). Quality Counts 2017: Under Construction—Building on ESSA's K-12 Foundation. *Education Week*.

Eisenhart, M., Weis, L., Allen, C. D., Cipollone, K., Stich, A., & Dominguez, R. (2015). High school opportunities for STEM: Comparing inclusive STEM-focused and comprehensive high schools in two US cities. *Journal of Research in Science Teaching*, *52*(6), 763–789. doi:10.1002/tea.21213

Ejiwale, J. (2003). Barriers To Successful Implementation of STEM Education. *Journal of Education and Learning*, *7*(2), 63–74.

Forman, J., Gubbins, E. J., Villanueva, M., Massicotte, C., Callahan, C., & Tofel-Grehl, C. (2015). National survey of STEM high schools' curricular and instructional strategies and practices. *NCSSS Journal*, *20*(1), 8–19.

Freel, A. (1998). Achievement in urban schools: What makes the difference? *Education Digest*, *64*(1), 17.

García, E., & Weiss, E. (2017). *Education Inequalities at the School Starting Gate: Gaps, Trends, and Strategies to Address Them*. Economic Policy Institute.

Gerber, B. L., Cavallo, M. L., & Edmund, A. M. (2001). Relationships among informal learning environments, teaching procedures and scientific reasoning ability. *International Journal of Sciences*, *23*(5), 535–549.

Gist, C. D., Bianco, M., & Lynn, M. (2018). Examining Grow Your Own Programs Across the Teacher Development Continuum: Mining Research on Teachers of Color and Nontraditional Educator Pipelines. *Journal of Teacher Education*. doi:10.1177/0022487118787504

Hansen, M., & Gonzalez, T. (2014). Investigating the Relationship between STEM Learning Principles and Student Achievement in Math and Science. *American Journal of Education*, *120*(2), 139–171. doi:10.1086/674376

Harper, N. W., & Daane, C. J. (1998). Causes and reduction of math anxiety in preservice elementary. *Action in Teacher Education*, *19*(4), 29–38. doi:10.1080/01626620.1998.10462889

Hibpshman, T. (2004). *A review of value-added models*. Frankfort, KY: Kentucky Education Professional Standards Board.

Hill, H. (2009). Fixing teacher professional development. *Phi Delta Kappan*, *90*(7), 470–476. doi:10.1177/003172170909000705

Hill, P. D., Friedland, E. S., & Phelps, S. (2012). How Teacher Candidates' Perceptions of Urabn Students Are Influenced by Field Experiences: A Review of the Literature. *Action in Teacher Education*, *34*(1), 77–96. doi:10.1080/01626620.2012.642290

Hird, M. D. (2013). *Barriers to implementation of new programs and pedagogies in K-12 STEM Education: a systems perspective*. Retrieved from http://hdl.handle.net/1721.1/81121

Hudley, C. (2013, October). Education and urban schools. *The SES Indicator, 6*(2).

Indiogine, S. P. (2015). The Achievement Gaps in Mathematics and Science. In A. Sahin (Ed.), *A Practice-based Model of STEM Teaching* (pp. 15–28). Rotterdam: SensePublishers. doi:10.1007/978-94-6300-019-2_2

Krishamurthi, A. (2014). *Examining the impact of afterschool STEM programs.* Academic Press.

Lewis, L., Snow, K., Farris, E., Smerdon, B., Cronen, S., & Kaplan, J. (1999). Condition of America's public school facilities. *Education Statistics Quarterly*, *2*, 42–46.

Lippman, L., Burns, S., & McArthur, E. (1996). *Urban Schools: The Challenge of Location and Poverty*. Nationa Center for Education Statistics.

Lloyd, S. D., & Harwin, A. (2016). Nation's Schools Get Middling Grade on Quality Counts Report Card. *Education Week*.

Matthew, R. (2017, May 10). *Students Shouldn't Live in STEM Deserts*. Retrieved from U.S. news and World Report: https://www.usnews.com/opinion/knowledge-bank/articles/2017-05-10/the-us-must-address-disparities-in-access-to-stem-education

McAnn, A. (2018, July 30). *States with the Best & Worst School Systems*. Retrieved from www.Wallethub.com: https://wallethub.com/edu/states-with-the-best-schools/5335/

Meguid, E. A., & Collins, M. (2017). Students' perceptions of lecturing approaches: traditional versus interactive teaching. *Advances in Medical Education and Practices*, 229-241.

Mervis, J. (2011, September 14). Is There a Special Formula for Successful STEM Schools? *Science Insider.* Retrieved from http://news.sciencemag.org/scienceinsider/2011/05/-is-there-a-special-formula-for-.html

Milner, H. R. IV. (2012, May 17). But What is Urban Education. *Urban Education*, *47*(3), 556–561. doi:10.1177/0042085912447516

Morgan, M., Farkas, G., Hillemeier, M. M., & Maczuga, S. (2016). Science Achievement Gaps Begin Very Early, Persist, and Are Largely Explained by Modifiable Factors. *Educational Researcher*, *45*(1), 18–35. doi:10.3102/0013189X16633182

National Center for Education Statistics. (2013). *Urban Education in America.* U.S. Department of Education. Retrieved from https://nces.ed.gov/surveys/urbaned/students.asp

National Commission on Teahing and America's Future. (1996). *What Matters most: Teaching for America's Future*. New York: Teachers College Columbia University.

National Research Council. (2011). *Successful K-12 STEM education: Identifying effective approaches in science, technology, engineering, and mathematics*. National Academies Press.

National Science Board. (2014). *Science and Engineering Indicators 2014*. Arlington, VA: National Science Foundation.

National Science Board. (2016). *Science and Engineering Indicators 2016*. Alexandria, VA: National Science Foundation.

Nwanekezi, A. U., & Nzokurum, J. C. (2010). Science teaching in Nigerian primary schools: The way forward. *African Journal of Education and Developmental Studies, 7*(1), 68-73.

Office of State Superintendent of Education. (2017). *2017 PARCC Results*. Office of State Superintendent of Education.

Office of the State Superintendent of Education. (2016). *2016 DC PAARC Results*. Washington, DC: Office of the State Superintendent of Education.

Onuja, J. E. (1987). *The causes of poor performance of secondary schools students in West African schools certificate Biology in Oyu L. G. A.* Ado-Ekiti University Library.

Organisation for Economic Co-operation and Development. (2018). *OECD Science, Technology and Industry Scoreboard*. Retrieved from OCED.org: http://www.oecd.org/sti/scoreboard.htm

PCAST. (2010). *Prepare and inspire: K-12 education in science, technology, engineering, and math (stem) for America's future.* PCAST.

Picardo, J. (2016, April 8). *A textbook problem: Seven suggestions to improve the quality of published resources; Exploring the cognitive principles that make a good textbook*. Retrieved from http://www.josepicardo.com/education/a-textbook-problem-seven-suggestions-to-improve-the-quality-of-published-resources/

President's Council of Advisors on Science and Technology. (2010). *Report to the President: Prepare and inspire: K-12 education in science, technology, engineering, and math (STEM) for America's future. Retrieved from.* National Academies Press. Retrieved from https://www.whitehouse.gov/sites/default/files/microsites/ostp/pcast-stemed-report.pdf

President's Council of Advisors on Science and Technology. (2012). Engage to excel: Producing one million additional college graduates with degrees in science, technology, engineering, and mathematics. Washington, DC: Author.

Programme for International Student Assessment (PISA). (n.d.). Retrieved from www.oecd.org

Rothwell, J. (2013). The Hidden STEM Economy. Washington, DC: Metropolitan Policy Program at Brookings.

Rothwell, J. (2014). *Still Searching: Job Vacancies ad STEM Skills*. Brookings Institute.

Schneider, B., Broda, M., Judy, J., & Burkander, K. (2014). Pathways to college and STEM careers: Enhancing the high school experience. *New Directions for Youth Development*, *140*, 9–29. PMID:24474256

Scott, A., & Martin, A. (2014). Perceived Barriers to Higher Education in STEM among High-Achieving. *Journal of Women and Minorities in Science and Engineering*, *20*(3), 235–256. doi:10.1615/JWomenMinorScienEng.2014006999

Seymour, E., & Hewitt, N. (1997). *Talking about leaving: Why undergraduates leave the sciences*. Boulder, CO: Westview.

Stuessy, N. E. (2015). Modeling Successful STEM High Schools in the United States: An Ecology Framework. *International Journal of Education in Mathematics, Science and Technology*, *3*(1), 77–92.

Subotnik, R., Tai, R., Rickoff, R., & Almarode, J. (2009). Specialized Public High Schools of Science, Mathematics, and Technology and the STEM Pipeline: What Do We Know Now and What Will We Know in 5 Years? *Roeper Review*, *32*(1), 7–16. doi:10.1080/02783190903386553

The After School Alliance. (2011). *STEM Learning in Afterschool: An Analysis of Impact and Outcomes*. Retrieved from http://www.afterschoolalliance.org/stem-afterschool-outcomes.pdf

The After School Alliance. (2013). *Defining Youth Outcomes for STEM Learning in Afterschool*. Author.

The Condition of Educaton: Concentration of Public School Students Eligible for Free or Reduced-Price Lunch. (2018). Institute of Education Sciences: National Center for Education Statistics.

Tirozzi, G., & Uro, G. (1997). Education reform in the United States: National policy in. *The American Psychologist, 52*(3), 241–249. doi:10.1037/0003-066X.52.3.241

Tooze, A. (2008). *The Wages of Destruction: The Making and Breaking of the Nazi Economy*. Penguin Books.

United States Government Accountability Office. (2012). *Science, Technology, Engineering, and Mathematics Education: Strategic Planning Needed to Better Manage Overlapping Programs Across Multiple Agencies*. United States Government Accountability Office.

Urban Education in America. (2013). Institute of Education Science: National Center Education Statistics.

U.S. Census Bureau. (2014, July 10). *Where do college graduates work?* Retrieved from United States Census Bureau: https://www.census.gov/dataviz/visualizations/stem/stem-html/

U.S. Department of Education. (2009). *Race to the Top Program Executive Summary*. Washington, DC: U.S. Department of Education.

U.S. Department of Education. (2016). *STEM 2026 A Vision for Innovation in STEM Education*. U.S. Department of Education.

U.S. Department of Education. (2016). *Summer Bridge Programs*. What Works Clearinghouse.

U.S. News/Raytheon STEM INdex. (2016). STEM Index. *U.S. News & World Report.*

Watson, D. (2011). What do you mean when you say Urban? *Rethinking Schools, 26*(1), 48–50.

Wei, R., Darling-Hammond, L., Andree, A., Richardson, N., & Orphanos, S. (2009). *Professional learning in the learning profession: A status report on teacher*. Dallas, TX: National Staff.

Weis, L., Eisenhart, M., Cipollone, K., Stich, A. E., Nikischer, A. B., Hanson, J., ... Dominguez, R. (2015). In the Guise of STEM Education Reform: Opportunity Structures and Outcomes in Inclusive STEM-Focused High Schools. *American Educational Research Journal*, *52*(6), 1024–1059. doi:10.3102/0002831215604045

Xie, Y., & Shauman, K. (2015). STEM Education. *Annual Review of Sociology*, *41*(1), 331–357. doi:10.1146/annurev-soc-071312-145659 PMID:26778893

Zucker, A. A., Shields, P. M., & Adelman, N. E. (1998). *A Report on the Evaluation fo the National Science Foundations's Statewide Systemic Initiatives (SSI)*. Program.

Chapter 2
STEM Career Interest at the Intersection of Attitude, Gender, Religion, and Urban Education

Philip R. Alsup
Maranatha Baptist University, USA

ABSTRACT

Inspiring learners toward career options available in STEM fields (science, technology, engineering, and mathematics) is important not only for economic development but also for maintaining creative thinking and innovation. Limited amounts of research in STEM education have focused on the population of students enrolled in religious and parochial schools in urban settings; yet given the historic conflict between religion and science, this large sector of American education is worthy of examination. This chapter incorporates Gottfredson's Theory of Circumscription and Compromise as it relates to occupational aspirations, Bem's Gender Schema Theory to explain the role of gender in career expectations, and Crenshaw's Intersectionality Theory as it pertains to religion and urban location as group identifiers. Practical interventions for encouraging young students to consider STEM careers are discussed.

DOI: 10.4018/978-1-5225-7814-7.ch002

INTRODUCTION

The prestige of American education among other developed nations of the world is at best tenuous. Several government, non-profit, and educational focus groups, such as the American Council on Education, the National Science Foundation, and the National Action Council for Minorities in Engineering agree that the United States is losing its footing in the global marketplace (Museus, Palmer, Davis, & Maramba, 2011). While the United States teeters on the edge of losing its global competitiveness, countries such as China and India report significant increases in college graduates in science, technology, engineering, and mathematics (STEM) fields (Wissehr, Concannon, & Barrow, 2011). Advances in STEM research and development abroad are causing many American companies to expand business overseas rather than domestically (Goldbrunner, Doz, Wilson, & Veldhoen, 2006; Malecki, 2010). While other nations of the world have reported improvement in STEM educational programs and have expanded research and development initiatives, America has been comparatively stagnant in these areas in recent years (National Science Board, 2014). For example, the National Science Foundation (2010) reported of the top 30 industrialized nations, students in the United States ranked 25th in math achievement and 21st in science achievement. The World Economic Forum (2009) made a comparable finding when ranking the quality of mathematics and science education in the United States as 48th in the world. Working to find solutions for this deficiency, some would suggest a stronger emphasis be placed on encouraging secondary students to pursue college degrees and careers in STEM fields (National Research Council, 2010; National Science Foundation, 2014). However, advancing this cause has grown increasingly difficult in the current field of K-12 education that is presently experiencing a dearth of STEM-degreed math and science teachers (Office of Postsecondary Education, 2013). While opportunities for employment in STEM fields appear to be growing, there is concern that fewer young adults are interested in pursuing STEM-related careers. For instance, in 2009, more Bachelor's degrees in America were awarded in business than in science, technology, engineering, and mathematics combined (National Center for Education Statistics, 2009). An alarming statistic presented by the U.S. Department of Education shows that as many as 65% of students who begin a STEM major do not finish within 6 years (National Center for Educational Statistics, 2009). More work is needed to understand why students are not interested in entering STEM fields and why those who are interested may not always complete degrees and enter the workforce as STEM professionals. Furthermore, a greater understanding of STEM education in a variety of learning environments may illuminate the discussion even further as educators seek methods for encouraging STEM pursuits in urban K-12 settings. Learning environments which cater to unique populations risk being under-studied and yet research in these populations is no less important (Tseng et al., 2013). Were researchers to focus attention on specific and unique K-12 learning environments, understanding of overall effective methods may be broadened.

BACKGROUND

The history of the American STEM emphasis can be traced back to the post-World War II era as the nation realized that research and scientific development had stagnated during wartime (Wissehr et al., 2011). A renewed emphasis on reviving innovation and developing new technologies culminated in 1950 with the establishment of the National Science Foundation, a distributor of funding to institutions engaging in scientific research. However, the government's interest in renewing a focus on science and research did not come solely from a realization that these pursuits had diminished during wartime. Other nations of the world were touting their own technological advances, and no achievement was more public than the launch of the Russian satellite *Sputnik* in 1957. Millions of Americans watched in awe as a foreign, communist power showcased its innovation, and many wondered what response the United States would have should Russian technology be used against American democracy. Hence, many viewed Russia's space superiority as a national security threat, realizing the potential for warheads to be delivered from across the globe. The nation quickly began to view science and technology as preservers of democracy, and a movement began to improve STEM education purely for the sake of self-preservation (Hein, 2006). The launch of *Sputnik* was a stark reminder of America's need to revitalize scholarly scientific research and to increase the rigor of education for young Americans. Ironically, then-Senator John F. Kennedy belittled Russia's rocketry pursuits, suggesting adventures into space were a waste of time and money (Dickson, 2001). However, within a few short years, Kennedy and other skeptics would come to view scientific innovation as the benchmark of global relevance.

The curricular and teaching reforms of the post-*Sputnik* era rightly placed strong emphasis on science as a system of thinking rather than as a body of knowledge to be memorized. The National Research Council (2011) in summarizing the view of science in the post-*Sputnik* era remarks that "Science, mathematics, engineering, and technology are cultural achievements that reflect people's humanity, power the economy, and constitute fundamental aspects of our lives as citizens, workers, consumers, and parents" (p. 2). Essentially, STEM education was not viewed as content to be learned only by top achievers but rather to be appreciated by all learners within the context of improving society and culture. Quality STEM education was and continues to be valuable because it creates good thinkers who employ observations in rational decision-making (Grover & Pea, 2013). In this perspective, STEM education does not merely produce smart workers but also well-rounded human beings who improve society and culture.

As the capstone of several decades of STEM focus, the National Research Council (2011) proposes greater effort be directed toward developing interests in STEM subjects in Elementary students with the overall goal of encouraging STEM-career paths. Echoing these recommendations, the National Academy of Sciences (2011) recommends bolstering confidence and interest for STEM-careers in young students and developing interventions which encourage children, especially females, to enter the STEM pipeline. As the fervency to increase the United States' STEM footprint intensifies, so too has the need to strengthen students' awareness of STEM career options.

Determining the age in which career choices are solidified is an area of discussion in current literature, and conclusions are mixed in regard to exactly when vocational ideas morph from dreams into serious goals. However, there is agreement among researchers that the age of career decisions is younger than commonly assumed (Bardick, Bernes, Magnussen, & Witco, 2008; Johnson, Ozogul, DiDonato, & Reisslein, 2013; Karatas, Micklos, & Bodner, 2011). The role adults play in helping children form career identities is also viewed in research as an important construct of vocational aspirations, and yet the influence of adult models, specifically in relationship to gender, is not fully understood (Redmond, et al., 2011; Sonnert, Sadler, & Michaels, 2013). No discussion of career interest would be complete without recognizing the role of attitude toward school subjects in the career interest of students. Literature reveals attitude and interest affect career choice and both can be influenced at an early age (Krogh, & Andersen, 2013; Rivera & Schaefer, 2009; Suri, & Sharma, 2013). Students' age, gender, attitudes toward school subjects, and perceptions of adult workers all contribute to the overall picture of understanding career interest.

There are several reasons for educators to be concerned with STEM education in general and with encouraging STEM-career paths specifically. While many professions and jobs in America may not require STEM-specific degrees, there are few sectors of the labor market that do not hinge on a working knowledge of STEM-related principles (Capobianco, Diefes-Dux, Mena, & Weller, 2011; Lacey & Wright, 2009). In some ways, interest in STEM is secondary to the necessity of developing the thinking abilities desired by society and the labor market. Basic literacy in and even technical understanding of STEM concepts are required for individuals to distinguish themselves and advance economically and professionally. Yet current students must recognize that scientists and engineers actually create jobs for others (Najafi et al., 2012). The National Academy of Sciences (2010) estimates that while scientists and engineers constitute just 4% of the labor market, an overwhelming majority of workers are employed because of the innovation and creativity of that group. The trickle-down effect of scientific research and creativity impacts the manufacturing, sales, communications, and transportation industries dealing with the logistics of invention and creativity (National Science Board, 2014). Young Americans have the opportunity to impact society on several levels if interest in STEM careers can be developed and maintained.

The nation's ability to grow economically depends on its citizens' abilities to innovate and think critically about scientific issues and contribute to ever-advancing technologies (National Academy of Sciences, 2010). In essence, the objective of increasing interest in STEM subject and career fields is not necessarily just to create more scientists but to increase the thinking ability of students in order that they become informed, rational thinking contributors of society (Wissehr et al., 2011). Relatively few students may become white-coat, laboratory scientists, and yet this stereotypical, unexciting identity is unfortunately often associated with scientific careers (Bennett & Hogarth, 2009; Christidou, 2011; Zhai, Jocz, & Tan, 2013). If more positive attitudes toward STEM subjects and careers can be developed, young people stand to view STEM careers more favorably and realize the importance of STEM in all occupations.

Research comparing rural and urban school settings has identified mixed outcomes in a variety of areas such as achievement (Logan & Burdick-Will, 2017), access to learning resources (Hardre & Sullivan, 2008; Wood & Howley, 2012), and teacher characteristics (Hardre & Sullivan, 2008). Additionally, within the context of STEM education, research suggests certain sectors of the American population respond uniquely to science, specifically those groups holding politically conservative views and espousing religious values (Fitzgerald & Glass, 2014; Longest & Smith, 2011; McCright & Dunlap, 2011; Mooney, 2006; 2012; Taber, 2013b). History documents a long-running conflict between religion and science, and this rift has been held liable for individuals of faith being reluctant to trust scientists or express interest in science-related careers (Baker, 2012; Billingsly, Taber, Riga, & Newdick, 2013; Evans, 2013). Specifically, conservative Protestants have been outspoken in their disagreements with modern scientific thoughts on topics such as evolutionary theory, climate change, and cloning (Coyne, 2012; Evans & Feng, 2013; Hmielowski, Feldman, Myers, Leiserowitz, & Maibach, 2013). This tension has resulted in religion being increasingly viewed as an unwelcome intrusion in science education (Taber, 2013b). However, despite the historic conflict between religion and science and current claims that conservative Protestants distrust scientists, relatively little research exists on the relationship between conservative Protestants and popular science (Bailey, 2012; Ceglie, 2013; Evans, 2013).

Current literature calls for interventions to support interest and positive attitudes toward science, technology, engineering, and mathematics and to encourage more school children to consider careers in those fields (Colvin et al., 2012; Hirsch, Berliner-Heyman, Cano, Kimmel, & Carpinelli, 2011; Kessels et al., 2014). The factors affecting children's aspirations are diverse, and yet literature on occupational development in STEM areas identifies several foundational, recurring constructs. As educators and researchers seek for practical ways to encourage positive STEM attitudes and career interest, there is a need to provide cost-effective interventions to promote these objectives across all levels of education. However, within the realm of STEM education, there is an unfortunate lack of empirical research specifically within the subset of Protestant Christian schools, the students of which make up a significant sector of American education. Conservative Protestants comprise over 26% of the American population (Pew Forum on Religion and Public Life, 2008) and therefore hold significant sway in national politics and movements. Further, conservative Protestant Christian schools tend to be found in higher-population areas giving these schools a significant voice in urban settings (Scifo, 2018).

Conservative Protestants have historically expressed skepticism of science and wariness toward scientists in general (Coyne, 2012; Gauchat, 2012; Li, 2013) and have even been identified as a "reservoir of skepticism" toward popular science and scientists (Evans & Feng, 2013, p. 597). If these dispositions about science and scientists are being conveyed to students enrolled in conservative, Protestant Christian schools, urban or otherwise, then research in STEM education can be enhanced through a better understanding of this demographic. This sizable subset is potentially in danger of being overlooked, and yet given its unique moorings and view of science, STEM education in Protestant Christian schools is worthy of closer examination.

THE INTERSECTING ISSUES

Theory of Circumscription and Compromise

The selection of an occupation is a process young people begin at an early age and may not actually finalize until well into adulthood. A person's self-concept of abilities, personality, and interests determines what vocations are viewed with potential and which are eliminated from consideration. Gottfredson's (1981) Theory of Conscription and Compromise informs the developmental process young people undergo as they advance through stages of vocational aspirations.

Gottfredson (1981) suggests a four-stage developmental theory of vocational aspirations to identify the cognitive processes young people engage with during the first two decades of life. In the first stage, *orientation to size and power* (ages 3-5), children begin to understand a realistic concept of adulthood and work—earning a living, being responsible, and making independent decisions. In the second stage, *orientation to sex roles* (ages 6-8), children develop conceptions of what society expects of males and females and the jobs that are typically held by members of each gender. In the third stage, *orientation to social valuation* (ages 9-13), young people develop ideas about how they fit into society's expectations by virtue of their abilities, intelligence, and gender. The final stage, *orientation to the internal, unique self* (ages 14 and older), brings the young adult into a realization of the opportunities available in the workforce and the role his or her interests, abilities, and social class play in choosing a career path.

During the latter stages of this development, young people engage in a process of eliminating certain vocational possibilities while retaining aspirations for others. Gottfredson (1981) exposes the need to understand individuals' self-concept as a significant factor of vocational goals, causing them to "seek jobs compatible with their images of themselves" (p. 546). However diverse the experiences of young people, some may never have opportunity to pursue career paths in which they may have enjoyed success simply due to lack of exposure to those vocations (Gottfredson, 2006). Stated another way, a significant constraint to vocational aspirations is ignorance about occupations compatible with a person's goals and interests. The challenge for educators according to Gottfredson (2006) is to "optimize experience, both by providing broad menus of vocationally-relevant activities for young people to sample and by promoting self-agency in seeking out formative experiences" (p. 168). The more vocational exposure and career education young people are provided, the better occupational decisions they may be equipped to make.

The Theory of Circumscription and Compromise brings together several constructs into a framework for understanding occupational aspirations. *Self-concept* is essentially a set of beliefs about who a person believes he or she is presently or could be in the future. Gottfredson (1981) suggests self-concept includes interests, known skills, and desire to fit into a particular social class. *Occupational images* are perceptions of occupations based on a person's knowledge of employees and job roles within a certain vocation. The perceptions individuals develop about occupations include generalizations of personality type, working conditions, gender, workload, compensation, and job responsibilities. These perceptions are a result of many influences, and as self-concept and occupational images are reconciled, the result is a long-term effect on openness or reluctance to entering certain careers.

Occupational preference then is a person's level of like or dislike for careers retained for consideration after self-concept and occupational images are weighed. This *social space* is the full set of perceived career possibilities the individual considers to have potential. Preferences may still include several improbable occupational goals, but the *accessibility of occupation* compels the individual to seriously consider potential obstacles for entry into certain fields. Economic, educational, social, and familial pressures may discourage continued consideration of certain fields, further narrowing viable career options. Gottfredson (2006) identifies this cognitive process as *circumscription* which is followed by a stage of *compromise* when the individual yields to more agreeable job alternatives. Gottfredson (1981) distinguishes the resultant *aspiration* as "the single occupation named as one's best alternative at any given time" (p. 548), and although the specific selection may change, it is typically drawn only from within the social space. In short, the Theory of Circumscription and Compromise describes a winnowing process whereby certain career options are eliminated, vocational alternatives are considered and reconsidered, and a career path based upon one's self-concept is identified and pursued. Gottfredson (1981) suggests the most significant factors impacting the process of circumscription and compromise are gender, social class, intellect, skills, and personal values. As a young person matures, certain of these factors play a larger role than others in narrowing the social space and determining which career aspirations should be eliminated and which should be seriously considered. The Theory of Circumscription and Compromise answers the question about which factors cause an adolescent to become narrow-minded in occupational aspirations, whereas just a few years prior, he or she was relatively open to many career ideas.

Gender-role identification is the first delimiting force a child reconciles with his or her perceptions of occupation, shying away from careers seen to be in conflict with his or her gender. Second, based on a perception of social class, children eliminate vocations lacking in prestige concomitant with their own social standing. Next, based on a self-concept of intelligence and ability, children will sideline jobs which would require significant effort in overcoming competence or aptitude deficiencies. Finally, by adolescence, young people begin to weigh more heavily their personal interests and values in the search for potential career paths. Thus, by adolescence, students have circumscribed a rather narrow space from which to draw remaining occupational options--they have in essence painted themselves into a corner of their choosing. As the time for actual job selection arrives, young people may find themselves in a position of compromising preference for availability, taking advantage of accessible job opportunities. However, Gottfredson (2005) points out the narrowing of career aspirations in childhood is largely based on social and economic perceptions of occupations and less on actual understanding of job descriptions, specific duties, and entry level requirements. In fact, while students may self-educate on careers of high interest, few may actually have a strong understanding of occupations they have eliminated from consideration (Gottfredson, 2005). Hence, the need to educate young people about job functions, requisite education, and day-to-day work duties is apparent in order for young people to make wiser, more informed vocational decisions.

Few middle school students are thoroughly knowledgeable about STEM vocational opportunities, and yet many young people narrow their career interests before full awareness has even taken place (Colvin, 2012; Johnson et al., 2013). Attitudes toward STEM subjects have been found to decrease rapidly during high school, further diminishing interests for STEM careers and possibilities (Lindner et al., 2012). Helping students value long-term goals and supplying information on steps needed to attain those aspirations may increase the number of students willing to remain open to the opportunities available in STEM fields (Koul et al., 2011). The Theory of Circumscription and Compromise suggests young students could potentially be influenced to refrain from eliminating STEM careers before an age of strong understanding has occurred.

In short, rather than attempting to *convince* students that STEM careers are the best choice, perhaps educators would be wise to simply focus on keeping younger students from *eliminating* STEM vocations from the range of potential career options. Remaining open to STEM-career options until greater maturity and real-world understanding has been reached would perhaps result in a greater number of young people pursuing careers in STEM fields.

Gender Schema Theory

An individual's cognitive construction of career choices is highly affected by his or her perception of gender roles. Exploring gender roles in a broad sense led Bem (1981) to develop Gender Schema Theory, spawning from the study of androgyny. Androgyny is described as "the presence of a high degree of desirable feminine and masculine characteristics in the same individual" (Santrock, 2007, p. 185), allegedly resulting in lower stress and better mental health (Bem, 1977). Gender Schema Theory suggests young people observe individuals in social contexts and construct a mental framework about what it means to be male or female (Bem, 1981). These self-constructed gender schemes shape the child's beliefs about acceptable gender behaviors, proficiencies, and future interests (Ormrod, 2006). The theory suggests that as a child encounters new knowledge and makes observations, he or she assimilates that information in terms of schemes about gender (Bem, 1981). In this way, society may purposefully or inadvertently teach children to *sex type* occupations and domestic roles.

Bem (1981) describes *sex typing* as the transmission of ideas about masculinity and femininity by surrounding culture, and the theory essentially holds society responsible for children's perceptions of what constitutes male and female. Thus, sex typing plays a role in the developing child's self-concept as he or she assimilates gender roles endorsed by society. Bem (1985) describes Gender Schema Theory as an amalgam of cognitive-developmental theory and social learning theory in that "sex typing is mediated by the child's own cognitive processing" while also "a learned phenomenon and hence…neither inevitable nor unmodifiable" (p. 186). The sex typed individual, according to Bem (1983), is in a sense handicapped by surrounding culture, and the more society transmits gender stereotypes to children, the greater the risk of spurning talent that otherwise might be tapped (Bem, 1984). As a resolution to this problem, Bem (1983) envisions a society that produces gender-aschematic children who are empowered to overcome sex typing and explore social roles free from the perceived constraint of gender. While separating gender completely from social roles may be improbable, Bem's (1993) desire to paint un-gendered horizons for children is an admirable, albeit lofty, goal.

Intersectionality Theory

The gender gap in STEM interest and vocations is well-established (Archer et al., 2013; Buse & Bilimoria, 2013; Liben & Coyle, 2014; Pearson & Miller, 2012; Rosser & Taylor, 2009; Suri & Sharma, 2013) and has drawn the attention of both private sector industry and the national government (National Research Council, 2011; Office of Postsecondary Education, 2013; National Science Foundation, 2014). Research suggests males tend to have more positive attitudes toward STEM subjects and higher interest in STEM careers. While Gender Schema Theory (Bem, 1981) informs the tendency of males and females to think differently about gender-appropriate vocations, the theory provides only a binary evaluation of the gender gap in science education. Therefore, the lens of an identity model employed in current research may provide additional clarity into the specific nature of female disinterest in STEM subjects and occupations.

Extending the general framework of Gender Schema Theory, Crenshaw (1989) popularized the concept of *intersectionality*, the notion that individuals are defined by multiple layers of group characteristics. According to Crenshaw (1989), overlapping traits such as race, ethnicity, and social class, should be considered when explaining disenfranchisement of certain groups. Much of Crenshaw's (1991) original work focused on a perceived lack of opportunity for African-American females, and Crenshaw articulated that although women in general suffered from discrimination, women of color were marginalized by society to a much greater extent than were

Caucasian women. Collins (2000) extended Crenshaw's work into a model now recognized as Intersectionality Theory: "the study of interlocking matrices of oppression" (Neitz, 2011, p. 59). Although the theory originated as an attempt to explain racial discrimination, Intersectionality Theory has since come to inform the hardships diverse groups in society experience in terms of intersecting and defining characteristics (Knudsen, 2006).

Essentially, Intersectionality Theory examines compounding disadvantages individuals face due to membership in specific groups. For example, Intersectionality Theory is currently informing perceived disadvantages to individuals in a number of areas such as special education (García, Ortiz, & Sorrells, 2012), politics (Doan & Haider-Markel, 2010; Foster, 2013; Levine-Rasky, 2011), sexuality (Collier, Bos, Merry, & Sandfort, 2013; Wadsworth, 2011; Warner, & Shields, 2013), and in the legal system (Best, Edelman, Krieger, & Eliason, 2011). Intersectionality Theory has also served as an interpretive lens for research on STEM pursuits (Bruning, Bystydzienski, & Eisenhart, 2012; Hanson, 2012, 2013), religion (Banton, 2011; Guittar, & Pals, 2014; Mirza, 2013; Schaeffer, & Mattis, 2012), Protestantism (Bulanda, 2011; Henward & MacGillivray, 2012; Neitz, 2013) and education in urban settings (Grant & Zwier, 2014). The research base in these areas provides an excellent theoretical backdrop in which to ground further research regarding career interest among students in unique learning populations.

Attitudes in STEM

Initiatives to increase positive attitudes toward STEM subjects have strengthened in recent decades and are now commonly included in state, federal, and private interest groups' curriculum standards (IPST, 2008). Increasing students' positive attitudes toward science is seen by some researchers as one of the primary responsibilities of the classroom teacher, and some suggest positive science attitudes have more important ramifications than does science achievement alone (Bennett & Hogarth, 2009; Cheung, 2007; Najafi et al., 2012). Educational reforms that help students maintain high attitudes toward STEM subjects may improve the likelihood of high-attitude students pursuing STEM-related careers, regardless of achievement level. Definitions of *attitude* and related constructs are abundant, and the construct of attitude can be a somewhat all-encompassing idea when used in everyday conversation. Yet authors tend to describe attitude in similar ways as the concept relates to education. Oskamp and Schultz (2005) describe attitude as the tendency for a person to respond positively or negatively toward a certain target based on prior thought and experience with that target. Essentially, a person responds to ideas, concepts, and in this context, school subjects, with conditioned thought patterns based on internalized opinions.

Kind, Jones, and Barmby (2007) discuss attitudes toward academic subjects in light of the strength of emotion with which a person reacts to an object due to beliefs about the object's worth or desirability. While some stimuli may provoke strong emotional responses, others with which the individual has less experience may cause minimal reaction. One of the most-cited and widely-accepted definitions of attitude in educational research comes from the work of Eagly and Chaiken (1993) who defined attitude as "a psychological tendency that is expressed by evaluating a particular entity with some degree of favor or disfavor" (p. 2). Reid (2011) suggests the most important aspect of the definition of attitude is the presence of a judgment on the part of the individual. Exactly where an individual's perception of a target rests on the continuum between favor and disfavor is the concept of *attitude*--a fluid concept research continuously attempts to quantify and describe. Understanding the internalized beliefs and ideas of students is important as these become the foundation for future decisions. Two well-established and related theories connect attitude with motivation. Ajzen and Fishbein's (1975, 1980) Theory of Reasoned Action holds that behaviors can be traced to attitudes formed in the past that created intentions in the individual to perform certain actions in the future. Students' intentions must be directed during youth if certain behaviors are to be anticipated in the future.

Ajzen (1985) refined the Theory of Reasoned Action and developed the Theory of Planned Behavior to first, understand constructs of prior attitude, subjective norms, and perceived behavior control, and second, to explain the decisions of individuals in light of past influences. Ajzen (1985) suggested an individual's goals for the future are subject to the amount of perceived control he or she has over factors relating to attaining those goals. In other words, individuals govern ideas and ambitions through an understanding of personal abilities and realistic expectations about factors under their control. The theories of Reasoned Action and Planned Behavior consider factors affecting decision making, providing a foundation for interventions in the present to produce desired outcomes in the future.

Fostering positive attitudes toward academic subjects can certainly increase enjoyment for school subjects and potentially increase achievement in the process. Interventions designed to increase enjoyment of and attitudes toward STEM subjects have generally been met with positive, and often long-term, results across various grade levels (Salta & Tzougraki, 2004; Wood, Ellison, Lim, & Periathiruvadi, 2011). Positive attitudes toward school subjects have been shown to be associated with strong motivation to do well in school, and researchers have argued for interventions and reforms to help teachers increase student motivation in the classroom (Koballa & Glynn, 2007; Koul, Lerdpornkulrat, & Chantara, 2011). Increasing motivation for school subjects not only raises achievement but also expands options and opportunities.

Interest in STEM

Interest, as a close relative of attitude, has been viewed as a leading factor of school motivation, even more so than personal beliefs of intelligence or competence (Inoue, 2007). Students who are motivated by interest in a subject are likely to persist in their study of that topic perhaps to a greater degree than those with stronger achievement (Diekman et al., 2010; Kerr et al., 2012; Ormrod, 2006). Mastery experiences in math and science may also cause students' interest levels and self-efficacy in those subject to be high, causing greater persistence in pursuing math and science in school and in career choices (Bøe, Henriksen, Lyons, & Schreiner, 2011; Usher & Pajares, 2009; Britner, 2008). Students, especially males, who establish strong interest in STEM subjects and careers during middle school and high school are more likely to pursue a college degree with STEM focus (Eccles, 2011; Morgan, Gelbgiser, Weeden, 2013). When viewed in the context of long-term possibilities, efforts spent during middle school and high school years on increasing interest for STEM subjects and careers can be a worthwhile investment (Dionne et al, 2011).

Interest can be described as the potential for a certain behavior to take place as the result of an attitude stance. Future actions cannot be guaranteed, and yet the willingness of the individual to pursue an action is considered his or her interest. The connection between attitude and interest is important due to the fact that while positive attitudes cannot perfectly predict behaviors, openness toward a certain behavior certainly precedes engaging in action (Glasman & Albarracin, 2006; Oskamp & Schultz, 2005). It follows that if positive attitudes toward STEM subjects and STEM careers can be nurtured during childhood, higher interest in pursuing further education and more openness to careers in these fields can be expected (Johnson et al., 2013; Kerr et al., 2012).

A student's willingness to find meaning in STEM subjects and his or her interest in pursuing a career in STEM may hinge on whether positive and authentic experiences were provided by educators (Karatas, et al., 2011; Kessels et al., 2014; Renninger, 2009). Unfortunately, middle school students often find science to be irrelevant to real-life and poorly explained by teachers (Barmby et al., 2008). While of course not representative of all science classrooms and students, these findings may unfortunately typify the experiences of all too many students. Conversely, students who appear to enjoy science and have positive attitudes toward studying science may not necessarily transfer that attitude into STEM-career interest (Dewitt et al., 2011). The precise connections between attitude, interest, and motivation provide abundant fodder for further educational research.

The National Research Council (2011) has suggested that more should be done during the early school years to expose students to the day-to-day work of STEM career professionals. In this recommendation, the NRC (2011) supposes students' perceiving scientists, mathematicians, engineers, and technology professionals to be normal, everyday people may improve attitudes toward these career paths. However, while more young people entering STEM career paths may be desirable and while more should be done to increase student awareness, interest alone is not a sufficient prerequisite for a successful STEM career (Kimmel, Miller, & Eccles, 2012). Commitment to further education, possession of talent, and skill development are also necessary to fulfill the demands of many STEM-related occupations. Those points conceded, students are not likely to commit themselves to requisite qualifications unless they have remained open to STEM-career choices in the first place.

Self-Efficacy and Student Motivation

Self-efficacy is often discussed alongside interest in the context of student motivation, and both play significant roles in school-subject goals and career aspirations (Inoue, 2007; Tuan et al., 2005a). Similar to interest, self-efficacy has also been found to increase aspirations for further study in science and in pursuing science careers (Dewitt et al., 2011). Researchers agree that self-efficacy coupled with interest is a very powerful combination for student success in academic pursuits (Dionne et al., 2011; Tuan et al., 2005a). However, self-efficacy alone may be an insufficient motivator if not combined with strong interest to cause actionable desire for pursuit of a target. Intrinsic desire to accomplish a goal can be a stronger motivator than feelings about the ability to be successful (Nosek et al., 2007), and therefore, encouraging students to explore interests regardless of efficacy level is an important responsibility educators (Cotabish, Dailey, Robinson, & Hughes, 2013).

Self-efficacy explains an individual's feelings of competence and includes four motivational sources impacting a student's belief about future success: 1) Mastery experiences—views about levels of previous performance, 2) Vicarious experiences—perceptions about the behaviors of others, 3) Social persuasion—opinions and viewpoints expressed by others, and 4) Physiological state—self-evaluation and interpretation of one's emotions (Bandura, 1997). Self-efficacy is an internal construct that may or may not be evidenced through resultant actions but one that may nonetheless give the individual a sense of confidence. Bolstering students' self-efficacy at an early age is highly important, and doing so before negative beliefs can take root is imperative (Bandura, 1997). Anxiety and negative feelings about school subjects can deter students from otherwise fulfilling careers in related occupations, and this is especially true in STEM-career fields (Bandura, 1986; Koul et al., 2011). Children develop emotional attachment or detachment for school subjects, and these feelings lead to higher interest and greater appreciation of learning's impact on the future (Glaser-Zikuda & Fusz, 2008). Equipping students with strong STEM-subject abilities is essential for academic confidence, but exposing students to the potentialities associated with that knowledge may be equally important.

STEM-Career Identity

Aspirations toward STEM careers can be enhanced through attention to STEM-subject attitudes and by building interest in STEM professions. However, unless students have an opportunity to realistically see themselves in those roles, chances for actually pursuing those jobs may be reduced (Bennett & Hogarth, 2009; Karatas et al., 2011). Research distinguishes this construct as *identity*—individuals' projections about what and who they would like to become (Eccles, 2009).

Entering into any profession, particularly a STEM field, involves a process through which young people construct an identity in light of experiences with school subjects and cognitive images of professionals in those fields (Christidou, 2011; Hoh, 2009; Karatas et al., 2011; Johnson et al., 2013). Krogh and Anderson (2013) describe science identity in terms of alienation from science or recruitment toward it—both internalized concepts of imagining oneself in the role of a STEM worker. For example, while a student may enjoy STEM subjects and express interest in exploring a STEM career, he or she may actually perceive scientists to be aging, white-haired men in lab coats (Farland-Smith, 2012). Negative perceptions of STEM professionals may unfortunately keep otherwise qualified students with positive science attitudes from developing a 'scientist' identity (Dewitt et al., 2011). Research studies utilizing methods such as "Draw a Scientist" often return uninspiring images of STEM professionals (Highfield, 2011; Zhai et al., 2013). For some students, especially females, identifying with STEM professionals is difficult due to the pervasive perception of scientists as an elite group of masculine intellectualists immersed in academia or a sterile laboratory (Archer et al., 2013; Carlone & Johnson, 2007; DiDonato et al., 2014). While this persona may actually be true for a small sector of STEM professionals, many scientists, mathematicians, and engineers are occupied in work that can be highly interesting and engaging. In fact, among students who have opportunities to experience the practical, hands-on work typical of STEM workers, perceptions of scientists as 'normal' increases (Covin et al., 2012; Welch & Huffman, 2011).

If children perceive STEM professionals in such a way that deters them from identifying with those careers, perhaps more should be done to promote positive, realistic imagery of STEM workers and the variety of jobs they hold. In two large studies of young students, both Archer et al. (2013) and Dewitt et al. (2011) found many children and parents are not knowledgeable about the diversity of jobs available in STEM fields. Understanding and evaluating the way in which identity construction impacts career choices is a valuable undertaking, and this is especially true in science education where future innovators and creative thinkers can potentially be developed (Kessels et al., 2014; Tucker-Raymond, Varelas, Pappas, Korzh, & Wentland, 2007).

In the study of identity, Giddens (1991) describes identity as a mental narrative a person continues to write about him or herself, adjusting one's ongoing autobiography as more experiences and interactions take place. Along the same lines, the Theory of Circumscription and Compromise informs the concept of identity as young people eliminate from their personal story certain narratives they view as incompatible with who they aspire to be (Gottfredson, 1981). Holland (2001) also approaches identity from a story-telling perspective but with the addition of an individual acting out a role he or she wants to project before others. Identity can be understood in some contexts as the way in which a person wants to be perceived by others and that which a person is willing to make known to peers, teachers, and family. This perspective sheds light on the emotional element of identity and the extent to which a person may cling to an aspiration. While some career ideas are loosely held, many vocational aspirations are only relinquished in the presence of insurmountable obstacles (Gottfredson, 1981).

Gender and STEM Interests

A person's family background, social circle, and awareness of culture all affect his or her projections of self in regard to work and education, yet few factors seem to be of equal or greater influence than the child's gender (Archer et al., 2013; Ayalon & Livneh, 2013; Eccles, 2011; Else-Quest, Hyde, & Linn, 2010; Kanny, Sax, & Riggers-Piehl, 2014; Koch, 2007). There are few studies in educational literature in which gender is not viewed as a variable of interest, and research in STEM education and STEM careers appears to be no exception (Eccles, 2007; 2011; Liben & Coyle, 2014; Stoeger et al., 2013). Methods for improving gender equity in STEM careers is debated in current literature, and there is a distinct disparity between males and females electing both to major in a STEM college program and then to pursue a STEM career thereafter (Hill, Corbett, & Rose, 2010; Miller & Kimmel, 2012; Smith, 2011; Smith, 2010; Watt, 2008). However, not all high school females decline in attitude toward science, and in some cases, females have shown to maintain or increase attitudes (Colvin et al., 2012; Cheung, 2007; Hirsch et al., 2011). For example, in a study of young females, Archer et al. (2013) found that although no significant difference was identified in *disinterest* toward science levels between boys and girls, significantly more boys than girls were found to be *keenly interested* in pursuing science. Although girls may not necessarily possess negative attitudes toward science at school, female interest in science certainly appears to wane beyond the classroom walls (Archer et al., 2013). School-age boys appear to maintain more positive attitudes toward science than girls (Brotman & Moore, 2008; Caleon & Subramaniam, 2008), and yet the cause for this disparity has largely been shown to be social forces rather than simply gender (Ceci, Williams, & Barnett, 2009). Females, for example, may view science as incompatible with aspirations for relationship-oriented careers or for jobs conducive to family life. These concerns tend to be quite important for many young women as they contemplate career choices (Diekman et al., 2010; Eccles, 2009; Frome et al., 2006; Spearman & Watt, 2013). Males, on the other hand, typically prioritize work and family values differently, and this tendency may result in a false association between gender *disparity* in STEM careers and female *disfavor* for STEM careers (Barmby et al., 2008). Whereas many girls imagine themselves in careers involving working with people, expressing creativity, and having time for family, few young girls believe STEM careers match these descriptors (Archer et al., 2013). Young women's persistence in medical fields such as nursing shows a preference for careers which encourage personal relationships and interaction with people, opportunities perhaps not always associated with other STEM professions (Eccles, 2011; Kimmel et al., 2012). Males report much stronger desire to work with 'things' and conduct investigations while females tend to prioritize social aspects of work (Su, Rounds, & Armstrong, 2009). Therefore, while many females may enjoy STEM subjects and be open to STEM careers, some may not necessarily pursue STEM careers more as a matter of practicality than as an issue of opportunity.

Although there is concern about low numbers of students aspiring to science-related careers, students typically have a high level of respect for STEM fields and vocations (Sjøberg, & Schreiner, 2007). Neither do wavering attitudes toward science and science careers appear to correlate to lack of aptitude for science or mathematics when accounting for gender (Haworth, Dale, & Plomin, 2008; Smith, 2011). Among engineering undergraduates, females tend to be more concerned with good grades and thus persist to attain high marks, whereas young men persist due to self-efficacy beliefs of accomplishing tasks (Concannon & Barrow, 2010). Boys may be less afraid of difficult courses and less concerned about grades, causing them to perhaps opt for more-challenging courses even at the risk of a reduced grade (Kimmel et al., 2012). Although males aspire to STEM professions more than females, boys may do so because they view themselves as suitable for STEM professions based on gender, not necessarily on achievement or valuation (Koul et al., 2011). High school students who have experienced good achievement in math and science are likely to enroll in advanced classes, but so too are students with high self-efficacy for math and science regardless of achievement (Eccles, 2009). Members of both sexes, while equally capable of achievement in STEM-related subjects, differ in desire to pursue those subjects, and thus a gender disparity may be due more to matters of motivation rather than ability.

One of the strongest influences in the minds of young people when making judgments about STEM careers is a perceived masculine gender association with STEM subjects and professions (Diekman, 2010; Hoh, 2009; Spearman & Watt, 2013). Science, math, and engineering are often viewed by females as masculine pursuits and not fitting within perceptions of feminine work (Ayalon & Livneh, 2013; Carlone & Johnson, 2007; Goldman & Penner, 2014). Children indicate perceptions of math and science as subjects boys enjoy more than girls and often label STEM subjects as masculine (Adamuti-Trache & Andres, 2008; Caleon & Subramaniam, 2008). Students generally report a belief that typical scientists are male, and females have reported a hesitancy to identify with scientists due to a view that science does not lend itself to interpersonal relationships or to the feminine persona (Buck, Plano-Clark, Leslie-Pelecky, Lu, & Cerda-Lizarraga, 2008; Sjøberg, & Schreiner, 2007). Even highly feminized STEM role models have been viewed as incongruent with typical femininity by middle school girls, indicating gender stereotypes in STEM fields may be difficult to overcome (Betz & Sekaquaptewa, 2012).

While gender differences in STEM attitude and achievement develop early, these differences become much more apparent in post-high school decisions. Students develop interests and form attitudes throughout their school experience, and young people in middle school and high school are especially in a stage of life in which plans and goals are being solidified (Gottfredson, 1981). School experiences in STEM subjects have a lasting effect on whether a student will pursue a STEM major in college, and high school achievement in STEM subjects can predict interest and persistence in a STEM major (Eccles, 2011; Shaw & Barbuti, 2010; Spearman & Watt, 2013). For example, achievement in second year Algebra has correlated highly with the likelihood of a student pursuing a STEM-related college degree (Ma & Johnson, 2008). However, among those choosing a STEM college major, females are less likely than males to actually finish a degree in a STEM field (Shaw & Barbuti, 2010). Even high-achieving females who finish college degrees in STEM areas sometimes lose desire to work in these fields early in a career, citing reasons such as raising children or dissatisfaction with long work hours (Diekman et al., 2010; Ferriman, Lubinski, & Benbow, 2009). Even females with strong aspirations for STEM careers in high school, college, and early adulthood are willing to relinquish this interest for the sake of achieving traditionally-held goals for women.

Many factors contribute to females becoming discontented with STEM professions. Yet oftentimes, females leave STEM careers, or fail to enter them, due to personal choices and preferences rather than lack of opportunity or gender discrimination (Ceci & Williams, 2010; Goulden, Frasch, & Mason, 2009). Increasing females' interest in STEM careers at a young age may be achievable, but convincing them to stay may be the greater challenge.

STEM Education in Urban Christian Schools

Science education in the public arena differs considerably from science education in Protestant Christian schools (Armstrong, 2010; Van Brummelen, 2009). Science in Christian education is not merely a "modification of the approach used in non-Christian schools…it is rather a radical reorientation of that topic within the philosophical framework of Christianity" (Knight, 2006, p. 236). Understanding the fundamental difference in philosophy between secular and Christian educational institutions is important for interpreting outcomes found in either population. Current research also suggests a large percentage of Christian schools are located in urban areas (Scifo, 2018), and hence, understanding STEM Education in Christian schools specifically may also broaden understanding on urban STEM education generally.

In Christian schools, science education builds upon a foundation that is believed to be established by a creator God through the world He created (general revelation) and His written communication with mankind (special revelation) known more formally as the Bible (Horton, 1992; Noebel & Edwards, 2002). Christian educators generally believe God established the natural world for man to enjoy, study, and derive insight into His character (Horton, 1992). Therefore, Christian science education holds that human scientific discoveries are not a final authority but must rather be interpreted in light of Biblical principles. Van Brummelen (2002) suggests this stance by Christian educators does not entail a wholesale rejection of secular science but rather that scientific conclusions must be tempered against a Biblically-based worldview of a depraved human nature.

Christian educators, for example, view Darwinian evolution as a theory developed within a worldview which seeks to exclude a creator God from reality (Knight, 2006). In practice, evolutionary theory has led publishers of secular science textbooks to "promote values at odds with those of Christianity" (Van Brummelen, 2002, p. 157). The outcome of an evolutionary view of origins is that mankind has the capability through science and ingenuity to improve the human race and eradicate social ills (Armstrong, 2010; Colson & Pearcey, 2004). Colson and Pearcey (2004) expose the irony that whereas secular science education boasts independence from religion, faith placed in science's potential to cure social problems may be just as strong a religious stance as the faith of Christianity. Knight (2006) reinforces this point by suggesting that public education produces students who place their "uncritical faith in science and mathematics rather than in the Creator of scientific and mathematical reality" (p. 238). One effect of a secular view of science may be that many Americans believe faith in God is not needed to interpret the world, and many have therefore come to expect science and religion to be in conflict (Armstrong, 2010). This alleged clash between science and religion is certainly a factor to consider when studying science education in schools with religious affiliations.

Despite a suspected conflict between science and religion discussed in the literature, relatively little current research has specifically evaluated the relationship between religion and STEM-career interest. The impact of religion on perceived veracity of concepts discussed in science classrooms has been examined (Billingsley, 2013; Hanley, Bennett, & Ratcliffe, 2014; Hoven, 2013), and studies relating religious views to acceptance or denial of Darwinian evolution are not uncommon (Astley & Francis, 2010; Baker, 2014). Recent research has speculated religion may play a role in student attitudes toward STEM subjects (Astley & Francis, 2010; Cho, 2012; Liben & Coyle, 2014), and yet little research examines religion's influence on STEM-career interest. Given the close relationship between attitude and interest, the expectation that religion may intersect with career interest, especially in STEM education, is certainly reasonable.

Intersectionality Theory has served as a frame of reference for understanding how overlapping characteristics—such as gender, location, or political ideology—may cause various people groups to think (Crenshaw, 1989). Recent studies in Intersectionality Theory have analyzed religion as a potential challenge to success in areas such as careers and politics (Arifeen & Gatrell, 2013; Fitzgerald & Glass, 2014; Haq, 2013; Wadsworth, 2011), and hence, Intersectionality Theory may provide clarity on any connection between religion-based education, gender, and urban settings within STEM education.

SOLUTIONS AND RECOMMENDATIONS

Children of various ages can be expected to respond differently to questions of career interest and goals as they cognitively balance desires with self-concept (Gottfredson, 1981). Whether students benefit more from interventions aimed at increasing positive attitudes toward STEM subjects and careers in elementary, middle, or high school has been a matter of debate. The National Science Board (NSB, 2010) has recommended greater emphasis especially on teacher training and professional development for elementary teachers with the goal of increasing STEM awareness and interest among students. Both the NSB (2010) and the National Research Council (2011) have argued specifically that STEM-related careers should be emphasized during the elementary and middle school years to increase the future candidate pool of STEM professionals. As with any discipline, thoroughly developing talent and ability in STEM subjects takes years, and therefore, channeling the curiosity of children toward STEM topics at an early age is vital (Capobianco et al., 2011; Colvin et al., 2012; Keeley, 2009; Pratt, 2007; Turner & Ireson, 2010). A high number of STEM professionals indicate their interest and talent for science-related jobs began in elementary school (Benbow, 2012; Maltese & Tai, 2010). Therefore, identifying young students with an affinity for STEM subjects and careers and then supporting interests with research-based interventions may be a practical way to increase the number of STEM professionals in the workforce in the future.

Natural curiosity is strong during early childhood, making this stage an ideal time to build a strong foundation for the relevance of STEM subjects and increase understanding of STEM careers. Although still viewing science as useful, students in late middle school and high school often report their desire to study science or pursue science-related careers has decreased significantly (Barmby et al., 2008; George, 2006). Secondary school students have already formed strong opinions about STEM subjects and careers, and therefore, effort may be better spent on interventions with younger students who are more fluid in their cognition (Johnson et al., 2013; Turner & Ireson, 2010). The school years between elementary and high school are a period in which attitudes toward science decline, and unfortunately, middle school students especially often eliminate STEM careers as options before they even know much about them (Lindner et al., 2012). Perhaps more can and should be done to present younger students with accurate views of real-life STEM professionals, breaking through false perceptions of scientists as predominantly aging, white-haired men working in a laboratory. The more that is done to impress students with the wide variety of enjoyable STEM-careers options, the greater the potential for young people to formulate positive visions of where STEM interest can lead.

Students in late-elementary and middle school are refining ideas about career goals and aspirations. Yet while young people are making life-long decisions about jobs and occupations, few actually have quality, personal experience with professionals in specific occupations, especially with those in STEM-related careers (Bardick et al., 2008; Chen & Cowie, 2014; Ware & Stein, 2013). This lack of exposure may cause students, especially females, to be disinterested in or unaware of STEM careers (Colvin, Lyden, & León de la Barra, 2012; Diekman, et al., 2010). Providing young students with practical experiences with professionals in certain careers may serve to strengthen interest in those careers and enable students to have more positive attitudes toward entering those professions. While there are calls for educational interventions to increase STEM-career interest, further research is needed to identify practical techniques for promoting this awareness (DiDonato, Johnson, & Reisslein, 2014; Iskander & Kapila, 2011). Educators would be wise to offer more research-based interventions during the elementary and middle school years to increase student interest in STEM-careers (Rivera & Schaefer, 2009). While many teachers may support an increased STEM focus, there is a need to provide curriculum materials that are easy to implement and not time-consuming (DiDonato et al., 2014; Stoeger, Duan, Schirner, Greindl, & Ziegler, 2013; Tseng, Chang, Lou, & Chen, 2013).

One such type of intervention has involved the use of videos of scientists to educate students on various STEM career options and to strengthen STEM attitude. Both long-term interventions using podcast interviews with STEM professionals (Wyss et al., 2012) and short-term interventions using multimedia presentations (Johnson et al., 2013) have been shown to increase students' interest in STEM careers. A qualitative study by Chen and Cowie (2013) used short video clips of scientists describing their work and responsibilities and then conducted interviews with a range of K-12 students. Student responses to the intervention were positive yet only anecdotal in nature. Ware and Stein (2013) examined the impact of video vignettes of scientists on high school students, seeking to differentiate between scientists as role models and scientists as mentors. The study provided recommendations for implementing videos in STEM classrooms, but little evaluation of STEM professionals' influence on attitude or interest was made.

Wyss et al. (2012) created videos of STEM professionals which were then shown to middle school students over a period of 8 weeks. The study revealed a statistically significant increase in STEM career interest among participants within the first 4 weeks, and yet the effect size was very small. Applying Intersectionality Theory, Michael and Alsup (2016) extended the work of Wyss et al. (2012) by creating a STEM-professionals interview video shown to middle school students enrolled in conservative, Protestant Christian schools. Michael and Alsup (2016) found a significant difference in attitudes toward both engineering and technology between male and female students, females having less favorable views of both. Future studies like these, implementing practical interventions before younger students, may be the key to solidifying STEM-career interest in the minds of children.

Current literature calls for interventions to address the loss of interest among students for STEM-subjects and careers, and there is much work to be done to grow the body of research related to these issues. Interventions that harness the curiosity in elementary years and bolster that openness during middle school may help reduce the rapid decline in STEM attitudes and career interests taking place before high school. Addressing the related constructs in light of Gottfredson's (1981) Theory of Circumscription and Compromise will increase understanding of the best time to focus attention on supporting career interest and knowing whether males and females can be expected to differ on expected outcomes. Gender Schema Theory and Intersectionality Theory provide a framework in which to consider the factors of urban location, gender, and religiously-based education simultaneously.

FUTURE RESEARCH DIRECTIONS

Students in elementary and middle school are refining ideas about career goals and aspirations (DeWitt et al., 2011; Lindner, et al., 2012; Wyss, et al., 2012). However, young people often lack experiential knowledge of potential career paths, especially those in the STEM arena (Bardick et al., 2008). Females may especially be disinterested in or unaware of STEM careers, keeping them from pursuing occupations for which they may actually possess potential (Diekman et al., 2010). Yet while the gender gap in STEM areas is fairly well-established, future research would do well to frame a better understanding of females in other unique learning environments. Religious and parochial school settings, private non-religious school settings, and urban and rural settings offer many opportunities to examine the effects of overlapping characteristics on the learning of young people. Research in these environments may prove beneficial and yet are underrepresented in the current literature base. Questions linger about the best age in which to introduce children to STEM careers, the influence of one defining characteristic (i.e. gender, setting, religion) over another, and the ability of curriculum design to influence STEM-career pursuits. Future research in these areas is warranted.

CONCLUSION

Young children often express interest in STEM career fields because of familiarity with those in the profession, and yet their knowledge of what a STEM career path actually entails is limited (Colvin, Lyden & Leon de la Barra, 2012; Johnson, Ozogul, DiDonato, & Reisslein, 2013; Karatas, Micklos, & Bodner, 2011; Ware & Stein, 2013). Attitudes toward STEM subjects and careers have been shown to decline during the late elementary and middle school years (Bennett & Hogarth, 2009; Boe, Henriksen, Lyons, & Schreiner, 2011), and yet interest in school subjects is closely tied to attitude toward those subjects and careers (Reid, 2011). By helping young students be more aware of STEM careers through unique interventions, more students may feel compelled to pursue STEM degrees and careers themselves. Several recommendations for classroom practice can be made relevant to this chapter's discussion. Practical experiences that expose young students to individuals in certain career paths have the potential to increase interest in those careers or at a minimum encourage positive attitudes toward those professions (Barmby et al., 2008; Hoh, 2009; Michael & Alsup, 2016; Wyss, 2012). Curriculum planning to include activities which portray STEM professionals as "real people" will promote a better understanding and perhaps greater desirability for STEM careers among K-12 students. Interventions such as in-person or virtual classroom visits by STEM professionals as well as student field trips to STEM-work environments may also bolster the perception that STEM careers very rarely involve sterile laboratories. By painting a realistic and exciting STEM-career picture in the minds of all K-12 students, the STEM landscape, especially in urban environments, might be improved in the years to come.

REFERENCES

Adamuti-Trache, M., & Andres, L. (2008). Embarking on and persisting in scientific fields of study: Cultural capital, gender, and curriculum along the science pipeline. *International Journal of Science Education, 30*(12), 1557–1584. doi:10.1080/09500690701324208

Ajzen, I. (1985). From intentions to actions: A theory of planned behavior. In J. Kuhl & J. Beckman (Eds.), *Action-control: From cognition to behavior* (pp. 11–39). Heidelberg, Germany: Springer-Verlag. doi:10.1007/978-3-642-69746-3_2

Ajzen, I., & Fishbein, M. (1980). *Understanding attitudes and predicting behavior*. Englewood Cliffs, NJ: Prentice Hall.

Archer, L., DeWitt, J., Osborne, J., Dillon, J., Willis, B., & Wong, B. (2013). 'Not girly, not sexy, not glamorous': Primary school girls' and parents' constructions of science aspirations. *Pedagogy, Culture & Society, 21*(1), 171–194. doi:10.1080/14681366.2012.748676

Arifeen, S. R., & Gatrell, C. (2013). A blind spot in organization studies: Gender with ethnicity, nationality and religion. *Gender in Management, 28*(3), 151–170. doi:10.1108/GM-01-2013-0008

Astley, J., & Francis, L. J. (2010). Promoting positive attitudes towards science and religion among sixth-form pupils: Dealing with Scientism and Creationism. *British Journal of Religious Education, 32*(3), 189–200. doi:10.1080/01416200.2010.498604

Ayalon, H., & Livneh, I. (2013). Educational standardization and gender differences in mathematics achievement: A comparative study. *Social Science Research, 42*(2), 432–445. doi:10.1016/j.ssresearch.2012.10.001 PMID:23347486

Baker, S. (2014). Faith-based schools and the creationism controversy: The importance of the meta-narrative. In *International handbook of learning, teaching and leading in faith-based schools* (pp. 177–192). New York, NY: Springer Netherlands. doi:10.1007/978-94-017-8972-1_9

Bandura, A. (1977). Self-efficacy: Toward a unifying theory of behavioral change. *Psychological Review, 84*(2), 191–215. doi:10.1037/0033-295X.84.2.191 PMID:847061

Bandura, A. (1977). *Social learning theory*. New York, NY: General Learning Press.

Bandura, A. (1986). *Social foundations of thought and action: A social cognitive theory*. Englewood Cliffs, NJ: Prentice Hall.

Bandura, A. (1997). *Self-efficacy: The exercise of control*. New York: W.H. Freeman.

Banton, M. (2011). Religion, faith, and intersectionality. *Ethnic and Racial Studies, 34*(7), 1248–1253. doi:10.1080/01419870.2011.582727

Bardick, A., Bernes, K., Magnusson, K., & Witko, K. (2008). Junior high school students' career plans for the future. *Journal of Career Development, 32*(3), 250–271. doi:10.1177/0894845305279168

Barmby, P., Kind, P., & Jones, K. (2008). Examining changing attitudes in secondary school science. *International Journal of Science Education, 30*(8), 1075–1093. doi:10.1080/09500690701344966

Bem, S. L. (1977). On the utility of alternative procedures for assessing psychological androgyny. *Journal of Consulting and Clinical Psychology, 45*(2), 196–205. doi:10.1037/0022-006X.45.2.196 PMID:850004

Bem, S. L. (1981). Gender schema theory: A cognitive account of sex typing. *Psychological Review, 88*(4), 354–364. doi:10.1037/0033-295X.88.4.354

Bem, S. L. (1983). Gender schema theory and its implications for child development: Raising gender-aschematic children in a gender-schematic society. *Signs (Chicago, Ill.), 8*(4), 598–616. doi:10.1086/493998

Bem, S. L. (1984). Androgyny and gender schema theory: A conceptual and empirical integration. In R.A. Dienstbier, & T.B. Sonderegger (Eds.), *Nebraska Symposium on Motivation* (Vol. 34). Lincoln, NE: University of Nebraska Press.

Bem, S. L. (1993). *The lenses of gender: Transforming the debate on sexual inequality*. New Haven, CT: Yale University Press.

Benbow, C. (2012). Identifying and nurturing future innovators in science, technology, engineering, and mathematics: A review of findings from the Study of Mathematically Precocious Youth. *Peabody Journal of Education, 87*(1), 16–25. doi:10.1080/0161956X.2012.642236

Bennett, J., & Hogarth, S. (2009). Would you want to talk to a scientist at a party? High school students' attitudes to school science and to science. *International Journal of Science Education, 31*(14), 1975–1998. doi:10.1080/09500690802425581

Best, R. K., Edelman, L. B., Krieger, L. H., & Eliason, S. R. (2011). Multiple disadvantages: An empirical test of intersectionality theory in EEO litigation. *Law & Society Review*, *45*(4), 991–1025. doi:10.1111/j.1540-5893.2011.00463.x

Betz, D. E., & Sekaquaptewa, D. (2012). My fair physicist? Feminine math and science role models demotivate young girls. *Social Psychological & Personality Science*, *3*(6), 738–746. doi:10.1177/1948550612440735

Billingsley, B. (2013). Students' perceptions of apparent contradictions between science and religion: Creation is only the beginning. In N. Mansour & R. Wegerif (Eds.), *Science education for diversity: Theory and practice* (pp. 329–338). New York, NY: Springer Netherlands. doi:10.1007/978-94-007-4563-6_16

Billingsley, B., Taber, K., Riga, F., & Newdick, H. (2013). Secondary school students' epistemic insight into the relationships between science and religion—A preliminary enquiry. *Research in Science Education*, *43*(4), 1715–1732. doi:10.100711165-012-9317-y

Bøe, M. V., Henriksen, E. K., Lyons, T., & Schreiner, C. (2011). Participation in science and technology: Young people's achievement-related choices in late-modern societies. *Studies in Science Education*, *47*(1), 37–72. doi:10.1080/03057267.2011.549621

Britner, S. L. (2008). Motivation in high school science students: A comparison of gender differences in life, physical, and earth science classes. *Journal of Research in Science Teaching*, *45*(8), 955–970. doi:10.1002/tea.20249

Brotman, J. S., & Moore, F. M. (2008). Girls and science. A review of four themes in the science education literature. *Journal of Research in Science Teaching*, *45*(9), 971–1002. doi:10.1002/tea.20241

Bruning, M., Bystydzienski, J., & Eisenhart, M. (2012). Intersectionality as a framework for understanding diverse young women's interest in engineering. *2012 WEPAN National Conference*, 1-14.

Buck, G. A., Plano Clark, V. L., Leslie-Pelecky, D., Lu, Y., & Cerda-Lizarraga, P. (2008). Examining the cognitive processes used by adolescent girls and women in identifying science role models: A feminist approach. *Science Education*, *92*(4), 688–707. doi:10.1002ce.20257

Bulanda, J. R. (2011). Doing family, doing gender, doing religion: Structured ambivalence and the religion-family connection. *Journal of Family Theory & Review*, *3*(3), 179–197. doi:10.1111/j.1756-2589.2011.00093.x

Buse, K., Bilimoria, D., & Perelli, S. (2013). Why they stay: Women persisting in US engineering careers. *Career Development International*, *18*(2), 139–154. doi:10.1108/CDI-11-2012-0108

Caleon, I. S., & Subramaniam, R. (2008). Attitudes towards science of intellectually gifted and mainstream upper primary students in Singapore. *Journal of Research in Science Teaching*, *45*(8), 940–954. doi:10.1002/tea.20250

Capobianco, B. M., Diefes-Dux, H. A., Mena, I., & Weller, J. (2011). What is an engineer? Implications of elementary school student conceptions for engineering education. *Journal of Engineering Education*, *100*(2), 304–328. doi:10.1002/j.2168-9830.2011.tb00015.x

Carlone, H. B., & Johnson, A. (2007). Understanding the science experiences of successful women of color: Science identity as an analytic lens. *Journal of Research in Science Teaching*, *44*(8), 1187–1218. doi:10.1002/tea.20237

Ceci, S. J., & Williams, W. M. (2010). Sex differences in math-intensive fields. *Current Directions in Psychological Science*, *19*(5), 275–279. doi:10.1177/0963721410383241 PMID:21152367

Ceci, S. J., Williams, W. M., & Barnett, S. M. (2009). Women's underrepresentation in science: Sociocultural and biological considerations. *Psychological Bulletin*, *135*(2), 218–261. doi:10.1037/a0014412 PMID:19254079

Chen, J., & Cowie, B. (2014). Scientists talking to students through videos. *International Journal of Science and Mathematics Education*, *12*(2), 445–465. doi:10.100710763-013-9415-y

Cheung, D. (2007). Students' attitudes toward chemistry lessons: The interaction effect between grade level and gender. *Research in Science Education*, *39*(1), 75–91. doi:10.100711165-007-9075-4

Cho, F. (2012). The limits of the Buddhist embrace of science. *Science and Engineering Ethics*, *18*(3), 539–542. doi:10.100711948-012-9361-z PMID:22426623

Christidou, V. (2011). Interest, attitudes and images related to science: Combining students' voices with the voices of school science, teachers, and popular science. *International Journal of Environmental and Science Education*, *6*(2), 141–159.

Collier, K. L., Bos, H. M., Merry, M. S., & Sandfort, T. G. (2013). Gender, ethnicity, religiosity, and same-sex sexual attraction and the acceptance of same-sex sexuality and gender non-conformity. *Sex Roles*, *68*(11-12), 724–737. doi:10.100711199-012-0135-5 PMID:23687403

Collins, P. H. (2000). It's all in the family: Intersections of gender, race, and nation. In U. Narayan & S. Harding (Eds.), *Decentering the Center: Philosophy for a Multicultural, Postcolonial, and Feminist World* (pp. 156–176). Bloomington, IN: Indiana University Press.

Colvin, W., Lyden, S., & León de la Barra, B. A. (2012). Attracting girls to civil engineering through hands-on activities that reveal the communal goals and values of the profession. *Leadership and Management in Engineering*, *13*(1), 35–41. doi:10.1061/(ASCE)LM.1943-5630.0000208

Concannon, J. P., & Barrow, L. H. (2010). Men's and women's intentions to persist in undergraduate engineering degree programs. *Journal of Science Education and Technology*, *19*(2), 133–145. doi:10.100710956-009-9187-x

Cotabish, A., Dailey, D., Robinson, A., & Hughes, G. (2013). The effects of a STEM intervention on elementary students' science knowledge and skills. *School Science and Mathematics*, *113*(5), 215–226. doi:10.1111sm.12023

Crenshaw, K. (1989). Demarginalizing the intersection of race and sex: A black feminist critique of antidiscrimination doctrine, feminist theory, and antiracist politics. *University of Chicago Legal Forum*, *1989*, 139–167.

Crenshaw, K. (1991). Mapping the margins: Intersectionality, identity politics, and violence against women of color. *Stanford Law Review*, *43*(6), 1241–1279. doi:10.2307/1229039

Dewitt, J., Osborne, J., Archer, L., Dillon, J., Willis, B., & Wong, B. (2011). Young children's aspirations in science: The unequivocal, the uncertain and the unthinkable. *International Journal of Science Education*, *33*, 1–50.

Dickson, P. (2001). *Sputnik: The shock of the century*. New York, NY: Walker Publishing.

DiDonato, M., Johnson, A., & Reisslein, M. (2014). A gender-specific, brochure-based intervention for improving boys' and girls' engineering stereotypes and academic self-perceptions. *Global Journal of Engineering Education*, *16*(1), 34–42.

Diekman, A. B., Brown, E. R., Johnston, A. M., & Clark, E. K. (2010). Seeking congruity between goals and roles a new look at why women opt out of science, technology, engineering, and mathematics careers. *Psychological Science, 21*(8), 1051–1057. doi:10.1177/0956797610377342 PMID:20631322

Dionne, L., Reis, G., Trudel, L., Guillet, G., Kleine, L., & Hancianu, C. (2011). Students' sources of motivation for participating in science fairs: An exploratory study within the Canada-Wide Science Fair 2008. *International Journal of Science and Mathematics Education, 10*(3), 669–693. doi:10.100710763-011-9318-8

Doan, A. E., & Haider-Markel, D. P. (2010). The role of intersectional stereotypes on evaluations of gay and lesbian political candidates. *Politics & Gender, 6*(1), 63–91. doi:10.1017/S1743923X09990511

Eagly, A. H., & Chaiken, S. (1993). *The psychology of attitudes.* London: Harcourt Brace Jovanovich College Publishers.

Earwood, E., & Suiter, P. (2012). *A scent of water: Bringing life back to the Christian school movement.* Greenville, SC: Ambassador International.

Eccles, J. (2009). Who am I and what am I going to do with my life? Personal and collective identities as motivators of action. *Educational Psychologist, 44*(2), 78–89. doi:10.1080/00461520902832368

Eccles, J. (2011). Gendered educational and occupational choices: Applying the Eccles et al. model of achievement-related choices. *International Journal of Behavioral Development, 35*(3), 195–201. doi:10.1177/0165025411398185

Eccles, J. S. (2007). Where are all the women? Gender differences in participation in physical science and engineering. In S. J. Ceci & W. M. Williams (Eds.), *Why aren't more women in science? Top researchers debate the evidence* (pp. 199–210). Washington, DC: American Psychological Association. doi:10.1037/11546-016

Else-Quest, N. M., Hyde, J. S., & Linn, M. C. (2010). Cross-national patterns of gender differences in mathematics: A meta-analysis. *Psychological Bulletin, 136*(1), 103–127. doi:10.1037/a0018053 PMID:20063928

Farland-Smith, D. (2012). Development and field test of the modified Draw-a-Scientist Test and the Draw-a-Scientist rubric. *School Science and Mathematics, 112*(2), 109–116. doi:10.1111/j.1949-8594.2011.00124.x

Ferriman, K., Lubinski, D., & Benbow, C. P. (2009). Work preferences, life values, and personal views of topmath/science graduate students and the profoundly gifted: Developmental changes and gender differences during emerging adulthood and parenthood. *Journal of Personality and Social Psychology*, *97*(3), 517–532. doi:10.1037/a0016030 PMID:19686005

Fishbein, M., & Ajzen, I. (1975). *Belief, attitude, intention and behavior: An introduction to theory and research.* Reading, MA: Addison-Wesley Publishing Co.

Fitzgerald, S. T., & Glass, J. L. (2014). *Religion and inequality in America: Research and theory on religion's role in stratification.* New York, NY: Cambridge University Press.

Foster, G. M. (2013). Faith and race in American political life. *The Journal of American History*, *99*(4), 1206–1207. doi:10.1093/jahist/jas594

Frome, P. M., Alfeld, C., Eccles, J., & Barber, B. L. (2006). Why don't they want a male-dominated job? An investigation of young women who changed their occupational aspirations. *Educational Research and Evaluation*, *12*(4), 359–372. doi:10.1080/13803610600765786

García, S. B., Ortiz, A. A., & Sorrells, A. M. (2012). Intersectionality as a framework for research and practice in special education. *Multiple Voices for Ethnically Diverse Exceptional Learners*, *13*(1), 1–3.

George, R. (2006). A cross-domain analysis of change in students' attitudes toward science and attitudes about the utility of science. *International Journal of Science Education*, *28*(6), 571–589. doi:10.1080/09500690500338755

Giddens, A. (1991). *Modernity and self-identity: Self and society in the late modern age.* Stanford, CA: Stanford University Press.

Glaser-Zikuda, M., & Fusz, S. (2008). Impact of teacher competencies on student emotions: A multi-method approach. *International Journal of Education and Research*, *47*(2), 136–147. doi:10.1016/j.ijer.2007.11.013

Glasman, L. R., & Albarracín, D. (2006). Forming attitudes that predict future behavior: A meta-analysis of the attitude-behavior relation. *Psychological Bulletin*, *132*(5), 778–822. doi:10.1037/0033-2909.132.5.778 PMID:16910754

Goldbrunner, T., Doz, Y., Wilson, K., & Veldhoen, S. (2006). The well-designed global R&D network. *Strategy and Business*, *43*, 15.

Goldman, A. D., & Penner, A. M. (2014). Exploring international gender differences in mathematics self-concept. *International Journal of Adolescence and Youth*, 1–16. PMID:27840545

Gottfredson, L. (2006). Circumscription and compromise. In J. H. Greenhaus & G. A. Callanan (Eds.), *Encyclopedia of career development* (pp. 167–169). Thousand Oaks, CA: Sage Publications. doi:10.4135/9781412952675.n62

Gottfredson, L. S. (1981). Circumscription and compromise: A developmental theory of occupational aspirations. *Journal of Counseling Psychology*, *28*(6), 545–579. doi:10.1037/0022-0167.28.6.545

Gottfredson, L. S. (2005). Applying Gottfredson's theory of circumscription and compromise in career guidance and counseling. In S. D. Brown & R. W. Lent (Eds.), *Career development and counseling: Putting theory and research to work* (pp. 71–100). Hoboken, NJ: John Wiley & Sons.

Goulden, M., Frasch, K., & Mason, M. A. (2009). *Staying competitive: Patching America's leaky pipeline in the sciences*. Center for American Progress and the Berkeley Center on Health, Economic, & Family Security. Retrieved November 20, 2013 from www.americanprogress.org/issues/2009/11/pdf/women_and_sciences.pdf

Grant, C. A., & Zwier, E. (Eds.). (2014). *Intersectionality and urban education: Identities, policies, spaces & power*. Charlotte, NC: Information Age Publishers.

Grover, S., & Pea, R. (2013). Computational thinking in K–12: A review of the state of the field. *Educational Researcher*, *42*(1), 38–43. doi:10.3102/0013189X12463051

Guittar, N. A., & Pals, H. (2014). Intersecting gender with race and religiosity: Do unique social categories explain attitudes toward homosexuality? *Current Sociology*, *62*(1), 41–62. doi:10.1177/0011392113516985

Hanley, P., Bennett, J., & Ratcliffe, M. (2014). The inter-relationship of science and religion: A typology of engagement. *International Journal of Science Education*, *36*(7), 1210–1229. doi:10.1080/09500693.2013.853897

Hanson, S. L. (2012). Science for all? The intersection of gender, race and science. *The International Journal of Science in Society*, *3*(2), 113–136. doi:10.18848/1836-6236/CGP/v03i02/51325

Hanson, S. L. (2013). STEM degrees and occupations among Latinos: An examination of race/ethic and gender variation. *Race, Gender, & Class*, *20*(1-2), 214–231.

Haq, R. (2013). Intersectionality of gender and others forms of identity: Dilemmas and challenges facing women in India. *Gender in Management*, *28*(3), 171–184. doi:10.1108/GM-01-2013-0010

Hardre, P. L., & Sullivan, D. W. (2008). Student differences and environment perceptions: How they contribute to student motivation in rural high schools. *Learning and Individual Differences*, *18*(4), 471–485. doi:10.1016/j.lindif.2007.11.010

Haworth, C. M., Dale, P., & Plomin, R. (2008). A twin study into the genetic and environmental influences on academic performance in science in nine-year-old boys and girls. *International Journal of Science Education*, *30*(8), 1003–1025. doi:10.1080/09500690701324190 PMID:20111670

Hein, G. (2006). Science education for a thriving democracy. *Hands On! 29*(1), 4–7, 19.

Henward, A., & MacGillivray, L. (2012). It's a combination of the Bible and what's in your heart: Unresolvable tensions and contested narratives in a southern child care center. *Journal of Curriculum Theorizing*, *28*(3), 91–108.

Highfield, R. (2011). Science presenter: Why is the popular image of scientists stuck in the past? *New Scientist*, *211*(2829), 29–30. doi:10.1016/S0262-4079(11)62210-0

Hill, C., Corbett, C., & St. Rose, A. (2010). *Why so few? Women in science, technology, engineering, and mathematics*. Washington, DC: American Association of University Women.

Hirsch, L. S., Berliner-Heyman, S., Cano, R., Kimmel, H., & Carpinelli, J. (2011, October). *Middle school girls' perceptions of engineers before and after a female only summer enrichment program*. Paper presented at the Frontiers in Education Conference, Rapid City, SD. 10.1109/FIE.2011.6142990

Hmielowski, J. D., Feldman, L., Myers, T. A., Leiserowitz, A., & Maibach, E. (2013). An attack on science? Media use, trust in scientists, and perceptions of global warming. *Public Understanding of Science (Bristol, England)*, *22*(3), 1–18. PMID:23825287

Hoh, Y. K. (2009). Using notable women in environmental engineering to dispel misperceptions of engineers. *International Journal of Environmental and Science Education*, *4*(2), 117–131.

Holland, D., Lachicotte, W., Skinner, D., & Cain, C. (2001). *Identity and agency in cultural worlds*. Cambridge, MA: Harvard University Press.

Hoven, M. (2013). A critical guide to resources in religion and science. *Religious Education (Chicago, Ill.)*, *108*(3), 280–297. doi:10.1080/00344087.2013.783361

Inoue, N. (2007). Why face a challenge? The reason behind intrinsically motivated students' spontaneous choice of challenging tasks. *Learning and Individual Differences*, *17*(3), 251–259. doi:10.1016/j.lindif.2007.02.002

Institute for the Promotion of Teaching Science and Technology (IPST). (2008). *National science curriculum standards: The basic education core curriculum B.E.25541*. Retrieved November 13, 2013, from ww.ipst.ac.th/web/images/files/SciCur_2008_EngVersion.pdf

Iskander, M., & Kapila, V. (2011). Revitalizing achievement by using instrumentation in science education (RAISE), a GK-12 fellows project. *Journal of Professional Issues in Engineering Education and Practice*, *138*(1), 62–72. doi:10.1061/(ASCE)EI.1943-5541.0000085

Johnson, A. M., Ozogul, G., DiDonato, M. D., & Reisslein, M. (2013). Engineering perceptions of female and male K-12 students: Effects of a multimedia overview on elementary, middle-, and high-school students. *European Journal of Engineering Education*, *38*(5), 519–531. doi:10.1080/03043797.2013.811477

Kanny, M. A., Sax, L. J., & Riggers-Piehl, T. A. (2014). Investigating forty years of STEM research: How explanations for the gender gap have evolved over time. *Journal of Women and Minorities in Science and Engineering*, *20*(2), 127–148. doi:10.1615/JWomenMinorScienEng.2014007246

Karatas, F. O., Micklos, A., & Bodner, G. M. (2011). Sixth-grade students' views of the nature of engineering and images of engineers. *Journal of Science Education and Technology*, *20*(2), 123–135. doi:10.100710956-010-9239-2

Keeley, P. (2009). Elementary science education in the K-12 System. *NSTA WebNews Digest*. Retrieved August 23, 2013 from http://www.nsta.org/publications/news/story.aspx?id=55954

Kerr, B., Multon, K., Syme, M., Fry, N., Owens, R., Hammond, M., & Robinson-Kurpius, S. (2012). Development of the distance from privilege measures: A tool for understanding the persistence of talented women in STEM. *Journal of Psychoeducational Assessment*, *30*(1), 88–102. doi:10.1177/0734282911428198

Kessels, U., Heyder, A., Latsch, M., & Hannover, B. (2014). How gender differences in academic engagement relate to students' gender identity. *Educational Research, 56*(2), 220–229. doi:10.1080/00131881.2014.898916

Kimmel, L., Miller, J., & Eccles, J. (2012). Do the paths to STEMM professions differ by gender? *Peabody Journal of Education, 87*(1), 92–113. doi:10.1080/0161956X.2012.642276

Kind, P., Jones, K., & Barmby, P. (2007). Developing attitudes towards science measures. *International Journal of Science Education, 29*(7), 871–893. doi:10.1080/09500690600909091

Knudsen, S. (2006). Intersectionality-A theoretical inspiration in the analysis of minority cultures and identities in textbooks. In E. Bruillar, B. Aamotsbakken, S. Knudzen, & M. Horsley (Eds.), Caught in the Web or Lost in the Textbook (pp. 61–76). Academic Press. Retrieved from http://www.iartem.no/documents/caught_in_the_web.pdf#page=61

Koballa, T. R., & Glynn, S. M. (2007). Attitudinal and motivational constructs in science learning. In S. K. Abell & N. Lederman (Eds.), *Handbook for research in science education* (pp. 75–102). Mahwah, NJ: Lawrence Erlbaum Associates, Inc.

Koch, J. (2007). A gender inclusive approach to science education. In D. Sadker & E. Silber (Eds.), *Gender in the classroom: Foundations, skills, methods, and strategies across the curriculum* (pp. 205–223). Mahwah, NJ: Lawrence Erlbaum Associates.

Koul, R., Lerdpornkulrat, T., & Chantara, S. (2011). Relationship between career aspirations and measures of motivation toward biology and physics, and the influence of gender. *Journal of Science Education and Technology, 20*(6), 761–770. doi:10.100710956-010-9269-9

Krogh, L., & Andersen, H. (2013). "Actually, I may be clever enough to do it". Using identity as a lens to investigate students' trajectories towards science and university. *Research in Science Education, 43*(2), 711–731. doi:10.100711165-012-9285-2

Lacey, T., & Wright, B. (2009). Employment outlook: 2008-18 Occupational employment projections to 2018. *Monthly Labor Review, 132*, 82.

Levine-Rasky, C. (2011). Intersectionality theory applied to whiteness and middle-classness. *Social Identities, 17*(2), 239–253. doi:10.1080/13504630.2011.558377

Liben, L. S., & Coyle, E. F. (2014). Developmental interventions to address the STEM gender gap: Exploring intended and unintended consequences. *Advances in Child Development and Behavior*, *47*, 77–115. doi:10.1016/bs.acdb.2014.06.001 PMID:25344994

Lindner, J. R., Rayfield, J., Briers, G., & Johnson, L. (2012). Graduate student fellowship program effects on attitude and interest toward science of middle school students. *Journal of Natural Resources and Life Sciences Education*, *41*(1), 15–21. doi:10.4195/jnrlse.2011.0021k

Logan, J. R., & Burdick-Will, J. (2017). School segregation and disparities in urban, suburban, and rural areas. *The Annals of the American Academy*, *674*(1), 199–216. doi:10.1177/0002716217733936 PMID:29430018

Ma, X., & Johnson, W. (2008). Mathematics as the critical filter: Curricular effects on gendered career choices. In H. M. G. Watt & J. S. Eccles (Eds.), *Gender and occupational outcomes: Longitudinal assessments of individual, social, and cultural influences* (pp. 55–83). Washington, DC: American Psychological Association. doi:10.1037/11706-002

Maltese, A. V., & Tai, R. H. (2010). Eyeballs in the fridge: Sources of early interest in science. *International Journal of Science Education*, *32*(5), 669–685. doi:10.1080/09500690902792385

Miller, J. D., & Kimmel, L. G. (2012). Pathways to a STEMM profession. *Peabody Journal of Education*, *87*(1), 26–45. doi:10.1080/0161956X.2012.642274

Mirza, H. S. (2013). A second skin: Embodied intersectionality, transnationalism, and narratives of identity and belonging among Muslim women in Britain. *Women's Studies International Forum*, *36*, 5–15. doi:10.1016/j.wsif.2012.10.012

Mooney, C. (2006). *The Republican War on Science*. New York, NY: Basic Books.

Mooney, C. (2012). *The Republican Brain: The Science of Why They Deny Science—and Reality*. Hoboken, NJ: John Wiley & Sons.

Morgan, S. L., Gelbgiser, D., & Weeden, K. A. (2013). Feeding the pipeline: Gender, occupational plans, and college major selection. *Social Science Research*, *42*(4), 989–1005. doi:10.1016/j.ssresearch.2013.03.008 PMID:23721669

Museus, S., Palmer, R. T., Davis, R. J., & Maramba, D. C. (2011). Racial and ethnic minority students' success in STEM education. *ASHE Higher Education Report*, *36*(6), 1–136.

Najafi, M., Ebrahimitabass, E., Dehghani, A., & Rezaei, M. (2012). Students' attitude toward science and technology. *Interdisciplinary Journal of Contemporary Research in Business*, *3*(10), 129–134.

National Academy of Sciences. (2010). *Rising Above the Gathering Storm, Revisited: Rapidly Approaching Category 5*. Washington, DC: The National Academies Press. Retrieved from http://www.uic.edu/home/ Chancellor/risingabove.pdf

National Academy of Sciences. (2011). *Joint G8+ science academies' statement on education for a science-based global development*. Washington, DC: The National Academies Press. Retrieved from http://www.nationalacademies.org/ includes/Final_Education.pdf

National Center for Education Statistics. (2008). *Characteristics of private schools in the United States: Results from the 2005-2006 private school universe survey* (NCES 2008-315). Retrieved November 6, 2013, from http://nces.ed.gov/pubs2008/pss0506/glossary.asp

National Center for Education Statistics. (2009). *Students who study science, technology, engineering, and mathematics (STEM) in postsecondary education.* Retrieved from http://nces.ed.gov/ pubs2009/2009161.pdf

National Center for Education Statistics. (2009). *Table 271: Bachelor's degrees conferred by degree-granting institutions, by field of study: Selected years, 1970-71 through 2007-08*. Retrieved from http://nces.ed.gov/programs/digest/d09/tables/dt09_271.asp?referrer=list

National Research Council. (2007). *Taking science to school: Learning and teaching science in grades K–8*. Washington, DC: National Academy Press. Retrieved from http://nap.edu/openbook.php?record_id=11625&page=1

National Research Council. (2011). *A framework for K–12 science education: Practices, crosscutting concepts, and core ideas*. Washington, DC: National Academy Press. Retrieved from http://nap.edu/openbook.php?record_id=13165&page=1

National Research Council. (2011). *Successful K-12 STEM Education: Identifying Effective Approaches in Science, Technology, Engineering, and Mathematics.* Washington, DC: The National Academies Press. Retrieved from http://www.stemreports.com/wp-content/uploads/2011/06/NRC_STEM_2.pdf

National Science Foundation, National Center for Science and Engineering Statistics. (2012). T*able 18: Doctorate recipients, by citizenship and broad field of study: Selected years, 1981–2011*. Retrieved from http://www.nsf.gov/statistics/sed/2011/pdf/tab18.pdf

National Science Foundation, National Science Board. (2010). *Preparing the next generation of STEM innovators: Identifying and developing our nation's human capital* (NSB-10-33). Retrieved September 19, 2013 from http://www.nsf.gov/nsb/publications/2010/ nsb1033.pdf

National Science Foundation, National Science Board. (2010). *Science and engineering indicators*. Washington, DC: The National Science Foundation. Retrieved from http://www.nsf.gov/statistics/seind10/pdf/seind10.pdf

National Science Foundation, National Science Board. (2014). *Science and engineering indicators 2014*. Washington, DC: The National Science Foundation. Retrieved from http://www.nsf.gov/statistics/seind14/content/overview/overview.pdf

Neitz, M. I. (2011). Feminist methodologies. In M. Stausberg & S. Engler (Eds.), *The Routledge Handbook of Research Methods in the Study of Religion* (pp. 54–67). New York, NY: Routledge.

Neitz, M. I. (2013). Insiders, outsiders, advocates and apostates and the religions they study: Location and the sociology of religion. *Critical Research on Religion, 1*(2), 129–140. doi:10.1177/2050303213490311

Noebel, D., & Edwards, C. (2002). *Thinking like a Christian: Understanding and living a biblical worldview*. Nashville, TN: B&H Publishing Group.

Nosek, B. A., Greenwald, A. G., & Banaji, M. R. (2007). The Implicit Association Test at age 7: A methodological and conceptual review. In J. A. Bargh (Ed.), *Social Psychology and the Unconscious: The Automaticity of Higher Mental Processes* (pp. 265–292). New York, NY: Psychology Press.

Ormrod, J. E. (2006). *Educational psychology: Developing learners*. Upper Saddle River, NJ: Pearson Education, Inc.

Oskamp, S., & Schultz, P. W. (2005). *Attitudes and opinions*. Mahwah, NJ: Lawrence Erlbaum Associates. doi:10.4324/9781410611963

Pearson, W. Jr, & Miller, J. D. (2012). Pathways to an engineering career. *Peabody Journal of Education, 87*(1), 46–61. doi:10.1080/0161956X.2012.642270

Pew Forum on Religion and Public Life. (2008). *U.S. religious landscape survey.* Washington, DC: Pew Research Center. Retrieved from: http://religions.pewforum.org/pdf/report-religious-landscape-study-full.pdf

Pratt, H. (2007). *Science education's "overlooked ingredient": Why the path to global competitiveness begins in elementary school.* Retrieved September 10, 2013 from http://science.nsta.org/nstaexpress/nstaexpress_2007_10_29_pratt.htm

Redmond, A., Thomas, J., High, K., Scott, M., Jordan, P., & Dockers, J. (2011). Enriching science and math through engineering. *School Science and Mathematics, 111*(8), 399–408. doi:10.1111/j.1949-8594.2011.00105.x

Reid, N. (2011). Attitude research in science education. In I. Saleh & M. Khine (Eds.), *Attitude research in science education: Classic and contemporary measurements* (pp. 3–44). Charlotte, NC: Information Age Publishing Inc.

Renninger, K. A. (2009). Interest and identity development in instruction: An inductive model. *Educational Psychologist, 44*(2), 105–118. doi:10.1080/00461520902832392

Rivera, L. M., & Schaefer, M. B. (2009). The career institute: A collaborative career development program for traditionally underserved secondary (6-12) school students. *Journal of Career Development, 35*(4), 406–426. doi:10.1177/0894845308327737

Rosser, S. V., & Taylor, M. Z. (2009). Why are we still worried about women in science? *Academe, 95*(3), 7–10.

Salta, K., & Tzougraki, C. (2004). Attitudes toward chemistry among 11th grade students in high schools in Greece. *Science Education, 88*(4), 535–547. doi:10.1002ce.10134

Santrock, J. W. (2007). *Adolescence.* New York, NY: McGraw-Hill Companies, Inc.

Schaeffer, C. B., & Mattis, J. S. (2012). Diversity, religiosity, and spirituality in the workplace. *Journal of Management, Spirituality & Religion, 9*(4), 317–333. doi:10.1080/14766086.2012.742750

Scifo, J. (2018). *An investigation of assessment, equal opportunity, and educational equity in conservative evangelical schools* (Doctoral dissertation). Retrieved from Liberty University Doctoral Dissertations and Projects. (1664)

Shaw, E. J., & Barbuti, S. (2010). Patterns of persistence in intended college major with a focus on STEM majors. *NACADA Journal, 30*(2), 19–34. doi:10.12930/0271-9517-30.2.19

Sjøberg, S., & Schreiner, C. (2007). Perceptions and images of science and science education. In M. Claessens (Ed.), *Communicating European Research 2005* (pp. 151–158). Springer Netherlands. doi:10.1007/1-4020-5358-4_26

Smith, E. (2010). Do we need more scientists? A long-term view of patterns of participation in UK undergraduate science programmes. *Cambridge Journal of Education*, *40*(3), 281–298. doi:10.1080/0305764X.2010.502886

Smith, E. (2011). Women into science and engineering? Gendered participation in higher education STEM subjects. *British Educational Research Journal*, *37*(6), 993–1014. doi:10.1080/01411926.2010.515019

Sonnert, G., Sadler, P., & Michaels, M. (2013). Gender Aspects of Participation, Support, and Success in a State Science Fair. *School Science and Mathematics*, *113*(3), 135–143. doi:10.1111sm.12007

Spearman, J., & Watt, H. M. (2013). Women's aspirations towards STEM careers. In W. Patton (Ed.), *Conceptualising Women's Working Lives* (pp. 175–191). Boston, MA: SensePublishers. doi:10.1007/978-94-6209-209-9_10

Stoeger, H., Duan, X., Schirner, S., Greindl, T., & Ziegler, A. (2013). The effectiveness of a one-year online mentoring program for girls in STEM. *Computers & Education*, *69*, 408–418. doi:10.1016/j.compedu.2013.07.032

Su, R., Rounds, J., & Armstrong, P. I. (2009). Men and things, women and people: A meta-analysis of sex differences in interests. *Psychological Bulletin*, *135*(6), 859–884. doi:10.1037/a0017364 PMID:19883140

Suri, G., & Sharma, S. (2013). The impact of gender on attitude towards computer technology and e-learning: An exploratory study of Punjab University, India. *International Journal of Engine Research*, *2*(2), 132–136.

Tseng, K. H., Chang, C. C., Lou, S. J., & Chen, W. P. (2013). Attitudes towards science, technology, engineering and mathematics (STEM) in a project-based learning (PjBL) environment. *International Journal of Technology and Design Education*, *23*(1), 87–102. doi:10.100710798-011-9160-x

Tuan, H.-L., Chin, C.-C., & Shieh, S.-H. (2005a). The development of a questionnaire to measure students' motivation towards science learning. *International Journal of Science Education*, *27*(6), 639–654. doi:10.1080/0950069042000323737

Tucker-Raymond, E., Varelas, M., Pappas, C., Korzh, A., & Wentland, A. (2007). "They probably aren't named Rachel": Young children's scientist identities as emergent multimodal narratives. *Cultural Studies of Science Education, 1*(3), 559–592. doi:10.100711422-006-9017-x

Turner, S., & Ireson, G. (2010). Fifteen pupils' positive approach to primary school science: When does it decline? *Educational Studies, 36*(2), 119–141. doi:10.1080/03055690903148662

Usher, E. L., & Pajares, F. (2009). Sources of self-efficacy in mathematics: A validation study. *Contemporary Educational Psychology, 34*(1), 89–101. doi:10.1016/j.cedpsych.2008.09.002

Wadsworth, N. D. (2011). Intersectionality in California's same-sex marriage battles: A complex proposition. *Political Research Quarterly, 64*(1), 200–216. doi:10.1177/1065912910376386

Ware, J., & Stein, S. (2013). From "Mentor" to "Role Model": Scaling the involvement of STEM professionals through role model videos. *Journal of Educational Multimedia and Hypermedia, 22*(2), 209–223.

Warner, L. R., & Shields, S. A. (2013). The intersections of sexuality, gender, and race: Identity research at the crossroads. *Sex Roles, 68*(11-12), 803–810. doi:10.100711199-013-0281-4

Watt, H. M. G. (2008). Gender and occupational outcomes: An introduction. In H. M. G. Watt & J. S. Eccles (Eds.), *Gender and occupational outcomes: Longitudinal assessments of individual, social, and cultural influences* (pp. 3–24). Washington, DC: American Psychological Association. doi:10.1037/11706-000

Welch, A., & Huffman, D. (2011). The effect of robotics competitions on high school students' attitudes toward science. *School Science and Mathematics, 111*(8), 416–424. doi:10.1111/j.1949-8594.2011.00107.x

Wissehr, C., Concannon, J., & Barrow, L. H. (2011). Looking back at the Sputnik era and its impact on science education. *School Science and Mathematics, 111*(7), 368–375. doi:10.1111/j.1949-8594.2011.00099.x

Wood, L., & Howley, A. (2012). Dividing at an early age: The hidden digital divide in Ohio elementary schools. *Learning, Media and Technology*, *37*(1), 20–39. doi:10.1080/17439884.2011.567991

World Economic Forum. (2009). *The Global Competitiveness Report 2009-2010.* Retrieved from http://www.weforum.org/pdf/GCR09/Report/Countries/United%20States.pdf

Wyss, V. L., Heulskamp, D., & Seibert, C. J. (2012). Increasing middle school student interest in STEM careers with videos of scientists. *International Journal of Environmental and Science Education*, *7*(4), 501–522.

Zhai, J., Jocz, J., & Tan, A. (2013). 'Am I like a scientist?': Primary children's images of doing science in school. *International Journal of Science Education*, *36*(4), 553–576. doi:10.1080/09500693.2013.791958

KEY TERMS AND DEFINITIONS

Attitude: Attitude toward STEM subjects describes a participant's favor or disfavor of science, technology, engineering, or mathematics as an academic pursuit.

Conservative Protestants: Individuals comprising a religious subgroup believing the Bible to be the ultimate source for truth claims about life and holding a literalist viewpoint of the origins of life as found in the book of Genesis.

Interest: The combination of emotional attachment and personal valuation of a task resulting in a desire for various levels of enjoyment.

STEM Career Interest: A participant's desire to pursue a career in a STEM field due to a perception of potential enjoyment.

STEM Professional: An individual who works in a science, engineering, technology or mathematical field and who holds a minimum of a Bachelor's degree in their professional field.

Section 2

Urban K–12 STEM Teacher Preparation and Professional Development

Chapter 3
Challenges to Implementing STEM Professional Development From an Ecological Systems Perspective

Zora M. Wolfe
https://orcid.org/0000-0001-6161-9235
Widener University, USA

ABSTRACT

This chapter provides insights to common barriers to conducting STEM professional development in an urban setting. The chapter will illustrate the complexity of integrating changes to STEM teaching practices from a systems perspective, from the lens of a researcher-practitioner conducting an 18-month study of professional development in a K-8 school in a metropolitan, public school setting. Specific challenges along themes related to the classroom level, the school level, and the district level will be discussed. Finally, the chapter concludes with implications and suggestions for future professional development providers, and schools and district leaders who are considering implementing instructional and curricular change related to STEM education.

DOI: 10.4018/978-1-5225-7814-7.ch003

INTRODUCTION

Educators conducting STEM professional development in urban settings often face significant challenges because urban schools are typically in large, densely populated areas with diverse populations, characterized by large enrollments and complex bureaucratic systems. The complexity of integrating changes to STEM teaching practices, from the lens of a researcher-practitioner conducting an 18-month study of professional development in a K-8 school in a metropolitan, public school setting will be illustrated within this chapter from a systems perspective. Using an ecological systems framework (Petrides & Guiney, 2002; Zhao & Frank, 2003), this chapter will provide a personal narrative and analysis of systemic barriers to change, to examine the challenges to impacting instructional practices and professional learning communities from the author's perspective as a researcher-practitioner who provided the professional development program within the school. An ecological systems framework provides a way to study the interactions within a school system, looking at relationships between what occurs in classrooms, with the structures and systems in place at the school and district level, and how they impact curricular and instructional change in the classroom (Zhao & Frank, 2003).

Successful implementation of the professional development program in this study was measured by the effectiveness of sustained professional development on teacher learning and instructional practices through the development of a STEM curriculum and implementing inquiry-based unit planning and teaching in an urban school. Underlying the foundation of the professional development was the expansion of the skills and knowledge within the staff to build effective professional learning communities (PLC) as teacher leaders, capable of inquiring into their practice to sustain and continue this STEM-focused improvement in their school community. PLCs, defined as "a group of people sharing and critically interrogating their practice," allow for teachers to engage in ongoing professional learning within their schools, enhancing student learning and school-wide change in practice (Stoll, Bolam, McMahon, Wallace, & Thomas, 2006).

Within the classroom level, the author will discuss themes related to teacher agency; teacher skills, knowledge, abilities, and existing beliefs and practices; response of students; and the impact of classroom resources. At the school level, themes related to leadership, scheduling, and staff turnover will be discussed. Finally, at the district level, the author will discuss the complexities of working within a large urban school district, specifically, themes related to navigating the district processes, integrating STEM curricular resources from the district level, and managing the number of competing initiatives mandated by the district. This chapter will conclude with implications and suggestions for future professional development providers, schools, and district leaders who are considering implementing instructional and curricular change related to STEM education.

BACKGROUND

The rationale for this study was based in research that documents challenges to implementing alternative pedagogies, specifically the use of inquiry-based teaching in elementary and middle school science classrooms (Abd-El-Khalick et al., 2004; Anderson, 2002; Capps & Crawford, 2013). There is also evidence that professional development of new curriculum and pedagogies requires continued support in order to be effectively implemented (Banilower, Heck, & Weiss, 2007; Desimone, Porter, Garet, Yoon, & Birman, 2002). However, because outside expertise is often a limited resource for urban schools, this study also intended to examine the creation of professional learning communities and the use of collaborative inquiry communities (a specific form of PLCs) within the process of developing a STEM curriculum and pedagogy, to build teacher self-efficacy to sustain new teaching practices beyond externally provided professional development.

The rationale for developing an interdisciplinary STEM curriculum was that an integrated STEM curriculum provides "opportunities for more relevant, less fragmented, and more stimulating experiences for learners'' (Furner & Kumar, 2007; p.186). In addition, an interdisciplinary curriculum has been shown to improve higher level thinking skills and problem solving, and improve retention (Ellis & Fouts, 2001; King & Wiseman, 2001; Smith & Karr-Kidwell, 2000). Also, integrating math and science has a positive impact on student motivation, interest in school, and achievement (Bragow, Gragow & Smith, 1995; Gutherie, Wigfield & VonSecker, 2000; Hurley, 2001). However, it is important to consider the professional development required to support teachers to develop, integrate, and implement an effective STEM curriculum.

According to the National Staff Development Council, "increasing the effectiveness of professional learning is the leverage point with the greatest potential for strengthening and refining the day-to-day performance of education" (Learning Forward, 2012, ¶3). Effective professional development can "help teachers deepen their knowledge and transform their teaching" (Borko, 2004). In particular, content-specific professional development has been shown to support positive changes in teaching practice (Garet, Porter, Desimone, Birman, & Yoon, 2001). Providing teachers with "opportunities to enact certain instructional strategies and to reflect, individually and collectively, on their experiences" during professional development programs, can support development of teachers' pedagogical content knowledge (PCK) specific to their subject area (Van Driel, & Berry, 2012).

It is recommended that professional learning in schools acknowledge known theories and principles of effective adult learning, aligning the needs of the learners along with the needs of the school (Zepeda, Parylo, & Bengtson, 2014). Specifically, standards for effective professional development should: include a concentrated focus on professional learning experiences that must be based on content derived from standards; include reflection, practice, and planning time for teachers; be comprised of activities that build inquiry skills and demonstrate modeling of these skills; highlight effective and consistent assessment activities; address concerns regarding changes in teaching, school and students; and be intensive and sustained (Parsons & Summers, 2004; Supovitz & Turner, 2000; Danter, 2005).

The use of professional learning communities as a form of professional development is one way to incorporate and encourage the types of practices that improve teachers' learning and instruction (DuFour & Eaker, 1998; Thompson, Gregg, & Niska, 2004). Teacher learning can occur within these professional learning communities when they have opportunities to inquire, collaborate, and reflect on their teaching practice (Cochran-Smith & Lytle, 1999; Darling-Hammond & McLaughlin, 1995).

Specifically, research shows that specific professional development related to curriculum adaptation and analysis, can lead to improving students' science learning (Penuel, Gallagher, & Moorthy, 2011; Roth, Garner, Chen, Lemmens, Schwille, & Wickler, 2011). As communities become more mature, Grossman and colleagues (2001) noted that collaboration and critical inquiry should be present: "Members begin to accept the obligations of community membership, which include the obligation to press for clarification of ideas and to help colleagues articulate developing understandings" (p. 990). Collaborative inquiry is defined as "a process consisting of repeated episodes of reflection and action through which a group of peers strives to answer a question of importance to them" (Bray, 2000, p. 6).

As such, collaborative inquiry provides a structure for professional development in schools and for improving teaching practice (Bray, 2002). Through this process, teachers have the opportunity to frame questions, examine evidence, reflect on practice, and make sense of their work in the classroom (Weinbaum, Allen, Blythe, Simon, Seidel, & Rubin, 2004). This process allows teachers to identify key problems or issues that are relevant to their specific context, so they can carry and sustain teaching reform in schools.

The research questions framing the study were 1) In what ways does ongoing professional development affect teachers' planning and implementation of an integrated STEM curriculum?, 2) What is the impact on teachers' instructional practices?, and 3) What is the impact on the professional learning communities within the school?

METHODS

Setting

This study was conducted at a K-8, Title I school in a large, metropolitan district, with approximately 600 students and a student-teacher ratio of 17:1, with 98% of the teachers having more than three years of teaching experience. The ethnic makeup of the school is 82% African American, 12% multiracial, 3% Asian, 2% Hispanic, and 1% White. In 2016-2017, approximately 25% of the students scored proficient or advanced in English Language Arts, and 8% of the students scored proficient or advanced in mathematics. Specifically, in science, overall, 32% of the students scored proficient or advanced, with 45% of the students scoring proficient or advanced on the Grade 4 science assessment, and 10% scoring proficient or advanced on the Grade 8 science assessment.

The researcher was initially approached by the principal of the school to provide STEM professional development for teachers since the school had recently received grant funding from the school district to engage in a school redesign program focused on STEM integration. Award of the grant was predicated on demonstration of staff and community commitment to the redesign process, and the grant provided funding specifically for professional development for the implementation of the redesign program.

In conjunction with the school leadership, a professional development program was designed based on the assessment of the existing science curriculum and teaching practices in the school, to support the development and implementation of a STEM focused curriculum. By creating a curricular foundation based on math and science as well as technology and engineering standards, the teachers would be able to utilize and integrate existing partnerships with structured and intentional strategies to engage students and support learning.

The professional development program was a structured workshop series for teachers to explore and engage in an understanding of STEM standards, practices, and teaching pedagogy, with an emphasis on developing curricular units that can be implemented in the upcoming school year. Specifically, the professional development consisted of weekly planning sessions with the school's teacher leadership team over the course of three months in the spring, to examine the scope and sequence of science instruction across the grades, and then a one-week workshop in the summer with teachers to future develop specific units of instruction utilizing the 5E Instructional Model (Bybee et al., 2006), and inquiry-based teaching practices. Over the course of the school year, two school-wide professional development sessions (one full day, and one half day) occurred to support the staff in revisiting their understanding of the 5E Instructional Model as they developed additional units, and monthly classroom visits and professional development sessions during grade level team meetings were conducted to observe and support teachers to implement the units.

Each curricular unit was designed with a focus on inquiry-based learning, which would have students "actively inquiring through thinking and doing…, often mirroring the processes used by scientists" (Minner, Levy, & Century, 2010). The National Research Council identified essential features of classroom inquiry as:

1. Learners are engaged by scientifically oriented questions.
2. Learners give priority to evidence, which allows them to develop and evaluate explanations that address scientifically oriented questions.
3. Learners formulate explanations from evidence to address scientifically oriented questions.
4. Learners evaluate their explanations in light of alternative explanations, particularly those reflecting scientific understanding
5. Learners communicate and justify their proposed explanations. (NRC, 2000, p. 25)

The 5E Instructional Model (Bybee et al., 2006) is a research-based constructivist model of teaching science that aligns well with the tenets of inquiry-based learning. The five E's (Engage, Explore, Explain, Elaborate, Evaluate) describe phases in an instructional sequence that are designed to provide coherent instruction to support the development of students' understanding of fundamental concepts in science. In the Engage phase, teachers access students' prior knowledge and use a series of activities to engage the learners in a new concept. In the Explore phase, teachers provide students with activities related to the concept to help them generate new ideas and explore the concepts, possibly by designing and conducting an investigation. During the Explain phase, teachers guide the students to develop explanations of the phenomena they've encountered to develop an understanding of the concepts being studied. The Elaborate phase allows students to extend their understanding by applying their newly developed knowledge to different experiences and activities.

Finally, in the Evaluate phase, students (and teachers) have the opportunity to assess their understanding of the new content. This instructional model differs from the traditional focus on vocabulary, memorization, and blinding following procedural directions to conduct an "experiment."

Underlying the foundation of this professional development was the expansion of the skills and knowledge within the staff to build effective professional learning communities (PLC) as teacher leaders, capable of inquiring into their practice to sustain and continue this improvement in their school community. To support this aspect of the work, the professional development sessions specifically included opportunities for teachers to reflect on, discuss, and practice skills necessary for PLCs to function, such as exploring issues of trust and vulnerability in sharing teaching practice, the utilization of protocols to examine work, and how to provide positive and critical feedback to peers.

Data

In the year of the study, the school employed 35 teachers, 18 of whom consented to participating in the study. All teachers whose teaching responsibilities included being the primary instructor for teaching science content were included. Teachers who were not included in the study included some special education teachers and other teachers who did not teach science (e.g. physical education teachers, music teachers, middle school teachers who only taught English and/or History). Of the 18 teachers, five were new to the school in the year the professional development occurred, but had experience teaching in other schools.

As a single-site case study, this study collected data from a variety of sources, including documents, direct observations, participant-observation, and interviews (Yin, 2003). Specifically, this study examined data collected during the professional development sessions including a needs assessment survey and teacher work samples that were created and discussed at the meetings. The needs assessment survey focused on the teachers' perception of their existing knowledge and comfort level with science teaching and the types of supports they would want to implement curricular change in science. During the professional development, teachers developed curricular materials (lesson plans, activities, assessments, projects, rubrics) for a STEM curriculum utilizing current best practices, which were included as documents in the data analysis. The curricular materials followed the district's unit planning template, with additional details related specifically to connections to the 5E's. These materials, and the work leading to the development of these materials (e.g. brainstorming posters, drafts of materials) were collected as part of the research study.

Direct classroom observations were conducted once a month over the course of the school year, where the researcher took detailed observational notes, focusing on instructional practices and content. Specifically, the researcher looked for evidence of activities and student and teacher behaviors aligned with characteristics of the phases of the 5E Instructional Model, as outlined in the Classroom Observation Inventory, developed by McHenry & Borger (2013).

Participant-observer data from the perspective of the researcher as a professional development provider also provided context for the study. The school-wide professional development sessions, as well as professional development sessions during grade level team meetings, were audiotaped to capture the exact words of the teachers for analysis as participant-observation data. A researcher journal with reflections after planning sessions with the school leadership team, classroom observations, professional development sessions, and interviews provided additional data for triangulation.

At the end of the school year, four teachers participated in a 45-60 minute semi-structured interview to describe their experience with the professional development. The four teachers who volunteered and consented to be interviewed represented a span of grade levels: kindergarten, fourth grade, seventh grade, and eighth grade. The interview questions were designed to elicit the teachers' perceptions of the professional development program, the development of their professional learning community, and the design and implementation of the curriculum. Examples of the interview questions include, "In what ways do you feel you are implementing good science teaching in your classroom?"; "In what ways, if any, have our professional development sessions helped you develop and implement your curriculum?"; and "How would you describe the way your learning community functioned during the course of this year?" Follow-up questions probed the teacher's perceptions of specific positive and challenging aspects of implementing their curriculum and making changes to their instructional practices.

The most pertinent data to providing insights to the common barriers to conducting STEM professional development in an urban setting included participant-observation data collected during professional development sessions and teacher interviews. Curricular artifacts and direct classroom observation data were used to triangulate and confirm patterns that were noted. The interviews and professional-development recordings were audio-taped and transcribed by the researcher. The transcripts were coded and analyzed to look for any corroborating themes related to the supports and challenges to collaborative inquiry. Using a pattern matching procedure (Yin, 2009), preliminary codes for indicators related to collaborative inquiry were developed initially. Then the codes were organized using the ecological systems framework to study the interactions within classrooms, schools, and the district (Zhao & Frank,

2003). Additional evidence from curricular artifacts and direct classroom observations were used to confirm the codes and themes. A few representative threads were read and coded with another researcher, and the coding scheme was refined to ensure reliability (Miles & Huberman, 1994, p. 64).

FINDINGS AND DISCUSSION

An ecological systems framework (Petrides & Guiney, 2002; Zhao & Frank, 2003) was used as a tool examine the challenges to impacting instructional practices and professional learning communities during the professional development provided in this study. Within the framework, themes relating to the classroom level, the school level, and the district level will be discussed.

Classroom Level

Within the classroom level, a number of challenges were noted related to teachers and students. Specifically, the analysis revealed themes related to teacher agency; teacher skills, knowledge, abilities, and existing beliefs and practices; response of students; and the impact of classroom resources.

Teacher Agency

With regards to teacher agency, the participant-observation data collected during professional development sessions revealed a relationship between how much agency teachers felt they had to implement change as to whether or not changes in instruction occurred. Teachers who felt empowered to make change were more likely to make changes in their curriculum and teaching than teachers who were unsure if they were "allowed" by the principal or district to make changes or teachers who continued to voice concerns about external obstacles, such as "I can't teach this [science content] because the district doesn't give us any books," or "We don't have money to buy supplies." Teachers who did not voice positive agency during the professional development sessions did not produce the amount of curricular materials who did voice agency, and did not plan as many science lessons for observation over the course of the year.

Teachers who successfully implemented the curriculum creatively utilized existing resources within the school, actively sought external resources, and were willing to experiment and try new teaching strategies that may have originally been uncomfortable, as was demonstrated in their classroom observations and during the planning time in professional development sessions. This type of willingness and ability to take action and make changes was present in the initial team of teachers who were involved in bringing forward the STEM learning grant but was not always necessarily the result of the formal leadership position on the team or specific authority or ownership given by the school leadership. In fact, one teacher voiced the attitude, "I might as well try something different because what we have been doing isn't working, so if it fails, it can't be any worse," during professional development sessions, further indicating that there was an internal willingness to take a risk, regardless of whether or not the leadership was supportive.

In this study, similar to other literature in the area of teacher agency (e.g. Calvert, 2016; Priestley, M., Edwards, Priestley, A., & Miller, 2012), the importance of fostering teacher agency was important in implementing change in STEM curriculum and teaching. Specific activities embedded within the professional development program related to building PLC and teacher leadership were included to build confidence and skills so that teachers feel a greater sense of agency. Empowering teachers to own their curriculum and adapt and change it to fit the needs of their students is an important step to facilitating curricular change. Specifically, school leaders can take steps to explicitly foster teacher agency and create a culture that allows teachers to experiment and take risks with trying new instructional strategies or activities.

Teacher Skills, Knowledge, Abilities, and Existing Beliefs and Practices

A teachers' skills, knowledge, abilities, and existing beliefs and practices relating to STEM teaching practices also influenced the impact of the professional development in curricular implementation during the school year. Unsurprisingly, teachers who initially noted a higher comfort level and/or interest in teaching math and science in the initial needs assessment survey were able to implement more of the curricular frameworks and pedagogical strategies that were discussed during the professional development sessions within their classroom observations. In several cases, these teachers had existing lessons and units they had previously taught and they were willing to examine how they could refine their lessons to utilize more inquiry-based approaches or how to integrate other subject areas, such as math and literacy standards and engineering practices.

For some teachers, a discomfort with engaging in science was apparent as they were asked to take on the role of the student in model lessons. During the large group professional development sessions, being unsure of the science content, or "the right answer", was a barrier to actively examining the structure of some units or trying new activities that were suggested in the district's curricular resources.

For many of these teachers, reframing the role of the teacher as a facilitator, rather than the expert, in inquiry-based teaching was reassuring.

However, for one teacher with a strong science content background, a deeper examination of science teaching pedagogy revealed gaps in how she was currently teaching, resulting in some resistance to shifting to a more inquiry-based teaching approach. For example, in her understanding of "correct science," it was difficult during the professional development sessions for her to recognize that pedagogical content knowledge is different than specific content knowledge, which created a barrier to recognizing the importance of acknowledging students' prior knowledge, which may include common pre-conceptions or misconceptions. Also, the teacher's focus on teaching the subject-specific content, prevented her from teaching the scientific practices; her focus in developing her curricular materials was on ensuring students knew definitions and could label diagrams, rather than being able to investigate or explain.

In recognition of the impact of existing skills, knowledge, abilities, and existing beliefs and practices, it is important for professional development providers and school leadership to assess for varying comfort levels with STEM education and differentiate professional development for teachers appropriately. As with other professional development, a "one-size-fits-all" approach could be detrimental to successful implementation of curricular change. Similar to how good teaching practice for students recognizes that preassessment and differentiation is important, there are also best practices for professional development that are also important to remember for adult learners as they learn more about STEM teaching and learning (e.g. Desimone & Garet, 2015).

Response of Students

Within the classroom level of analysis, the response of students to the new curricular changes and teaching approaches also impacted the amount of influence the professional development program had on the implementation of a new STEM curriculum. If teachers reported that their students were engaged and interested in the new lessons that were initially implemented, these teachers were more likely to continue developing new units and lessons during the professional development sessions. If initial implementation was not well received by students, teachers were less likely to be committed to continuing to develop new units or try new strategies, even if the teachers were initially enthusiastic about the new curriculum initiative. After implementing the initial lessons, some teachers made statements during the professional development conversations, such as, "You see, our students are not ready for this, yet," or, "It's not worth putting in all this time if the students still don't get it." However, for some teachers, statements such as, "Wow- I've never seen them [students] that engaged before," and "I didn't think they [students] would be able to figure it out, but they did!" indicated an excitement to stay engaged in the professional development and keep working toward curricular changes.

Because these initial student responses played a large role in whether or not teachers continued developing curriculum beyond the initial intensive summer workshop sessions, it is critical that adequate support is provided to develop initial lessons and units that are engaging to students and model what successful STEM teaching could look like. Ensuring that teachers start with a relevant, unique or high interest topic for students, as well as a lesson that teachers feel prepared and confident in teaching could positively impact continued teacher buy-in and long-term teacher engagement in STEM professional development. In addition, opportunities to examine student work related to the lessons, were helpful in engaging teachers during professional development meetings, because they were able to see where and how students were learning and engaging in the content.

Impact of Classroom Resources

As with many urban schools, a lack of classroom resources was noted by the teachers in this study in the needs assessments and their interviews, as a barrier to implementing curricular change in this study. As a participant-observer, an initial inventory of science materials indicated that some science teaching kits from a previous science initiative were available, but a number of kits were unable to be located and many consumable materials for activities and experiments within existing kits had not be replenished. Basic laboratory equipment, such as microscopes, thermometers, and safety goggles, was missing or in disrepair. The elementary grades (K-5) did not have a dedicated equipment storage area; rather, teachers kept science equipment and supplies, if they had any, within their individual classrooms. The middle school grades (6-8) had dedicated science classrooms, however, due to teacher turnover, middle school science equipment and supplies were also not sufficient or organized.

To support STEM curricular change, the curriculum team in the building spent a considerable amount of time gathering, creating an inventory, and organizing available science materials so that necessary materials could be ordered. Providing concrete suggestions to shift teacher mindset towards using available and more easily accessible materials was crucial to remove a perceived barrier that the school "had no science supplies" so therefore it was "not possible to do science." Referring teachers to activities that utilized items that could be purchased at hardware stores, office supply stores, pharmacies, or grocery stores, shifted the idea that "doing science" required test tubes, chemicals, or other expensive equipment that was unavailable.

In addition, a "science lab" space was created within the school for elementary teachers to utilize with their students in case planned activities may be "too messy" or require more space than was available in their classrooms. This space was also used to store science materials, so they could be shared and more easily accessible to all teachers, rather than limited to one individual classroom.

School Level

Continuing in the ecological framework, at the school level, themes related to leadership, scheduling, and staff turnover, were present in the data. School level themes included challenges that were present within the building, beyond the influence of individual classroom teachers, but were local to the specific school.

Leadership

Although the initial STEM school redesign initiative had the support of the principal and curriculum leadership team of the school, a number of factors related to leadership impacted the effectiveness of the professional development. Specifically, within this case study, the leadership of the principal was one factor that impacted the implementation of the professional development. As a school leader, the principal is responsible for creating and developing the mission and vision of the school (National Policy Board for Educational Administration, 2015). Ensuring that this STEM initiative aligned with the overall mission and vision of the school and that this vision is shared by all stakeholders is critical to the success of the initiative. In this case, the researcher discovered that although the principal and members of the curriculum leadership team were excited about the initiative, an understanding of the vision was not shared by all teachers, as apparent in the needs assessment survey, professional development sessions, and interviews. The rationale and plans for this initiative were not appropriately communicated across all grade levels and all teachers, resulting in some teachers being surprised that they were being asked to engage in this type of professional development. This created a struggle with developing truly collaborative professional learning communities because of the lack of buy-in and engagement.

In addition, clear communication from the principal about the roles, expectations, and responsibilities for the researcher, the principal, and the teachers could have facilitated better implementation of the professional development. In conversations with the teachers, they were often unclear about what they "had to do" in the classroom as a result of engaging in the professional development workshop. Although the specific objectives and outcomes for the professional development plan was to develop STEM lessons and units, accountability for implementing the plans in the classroom was something that needed to be supported by the principal since the researcher did not have the authority to require teachers to teach specific lessons.

Scheduling

Related to leadership within the school, the leadership of the school was responsible for determining the instructional schedule, which also had an impact on the implementation of the STEM professional development program. The instructional scheduled determined when teachers were being asked to teach science, when professional development sessions could be offered, and when teachers could plan together collaboratively.

It is important to note that within this school's schedule, there were opportunities for grade level planning, and specific time within the instructional schedule to teach science, both of which were helpful to implementing the STEM curriculum. However, in this school, the overall instructional schedule was changed several times over the course of the school year to accommodate changes in staffing. These shifts in the scheduled planning periods for the different grade levels resulted in confusion and difficulty in scheduling meeting times for planning PLC meetings and opportunities for classroom visits with the researcher. Within the overall school schedule, there were also very limited opportunities throughout the school year to meet with the entire faculty at the same time since there were no scheduled times for staff meetings beyond specific district-wide professional development days.

Also, all teachers were expected to teach science for 30 minutes at the end of day. Other than being logistically difficult to observe different science lessons during school visits, it was also difficult for teachers to share equipment and space since the whole school was supposed to engage in science at the same time. Teachers noted during the professional development sessions that the 30-minute time block also created the barriers of trying to fit a lesson into that time slot when often activities might require a longer amount of time. Teachers also shared that science being scheduled at the end of every day also meant that sometimes science wasn't taught because teachers "ran out of time" at the end of the day, or that students and teachers felt rushed to "cover" the material when energy levels and engagement were not at peak levels.

These challenges highlight the importance of creating a time for STEM professional development work to happen. For a professional development program to be effective, teachers need opportunities to work together and time to meaningfully engage in discussions and planning. In this school, when it was determined that it was difficult to make significant progress during the 30-minute team planning time within the schedule, an alternate plan was implemented to increase the time available for this professional development work by scheduling a half-day meeting for each grade level team to work with the researcher to begin to develop units and lessons for the upcoming quarter. In order for this to happen within the existing instructional schedule, the school invested in providing substitute coverage for the teachers to allow for these blocks of planning time, but the one longer block of sustained time allowed for more progress to occur within the same total number of minutes over multiple days.

Staff Turnover

As with many urban schools, staff turnover was a significant challenge in implementing curricular change through professional development. Changes in teaching staff made it difficult to sustain the vision of the STEM initiative; in this case, all of the teachers who were part of the initial development of the STEM school redesign grant application had either resigned or were transferred to other schools in the district by the beginning of the following school year. This created a lack of continuity and historical knowledge in understanding the rationale and original vision of the grant. In addition, a number of the teachers involved in completing the initial assessment of science instruction within the school and developing the scope and sequence were also no longer in their positions. Much of the initial work needed to be revisited and reviewed in the summer workshops since there were a number of new teachers, and relationships and trust needed to be re-established in new PLC groups. However, this turnover in staff did allow the leadership to hire new teachers who were specifically interested in STEM education.

In addition to teacher turnover, this school also experienced a change in administrative leadership during the school year. The school year began with a new assistant principal, who was not part of the original STEM school redesign plan, but then mid-year, the principal was reassigned, and the assistant principal become the leader of the STEM initiative when interim leadership was put in place. Without the support of the initial principal and because of the uncertainty of the direction that the school would take in the next school year under new leadership, teachers were unsure how to proceed. Several teachers voiced similar concerns to, “I don’t want to invest time into re-doing all of this curriculum if the new principal decides to take us in a completely different direction,” throughout professional development time and in individual interviews.

Although staff turnover is not surprising in urban schools, significant staffing changes are a challenge to overcome if implementing STEM professional development for curricular change. School and district leaders need to be committed to STEM initiatives and recognize the importance of retaining key personnel. In addition, broadening the ownership of the vision across the range of stakeholders can allow the initiative to continue even if there is some turnover, because there would continue to be other people who can move the vision forward.

District Level

In addition to the classroom and school level challenges, the ecological framework can be expanded to examine challenges at the district level. At the district level, the complexities of working within a large urban school district were revealed. Specifically, themes were found relating to navigating the district processes, integrating STEM curricular resources from the district level, and managing the number of competing initiatives mandated by the district.

Navigating the District Processes

In large urban districts, there are typically a number of different levels of leadership and a variety of rules and procedures that need to be followed. As part of implementing STEM professional development, as with other school initiatives, it is important for the people involved in the professional development processes to be comfortable and nimble in navigating the district systems and for the systems to be aligned with the professional development ideas.

For example, in this district, before the researcher could begin the research-practitioner based research study, the district required completion of a research orientation class prior to submitting a proposal to the district's external research review board. In addition, a new vendor application needed to be completed and approved, and a contract for the scope of work needed to be reviewed by a number of different offices. Obviously, these processes take time, so having prior familiarity with the systems or providing clear information for the professional development provider and/or principal would be helpful to efficiently navigate through the bureaucracy.

With respect to instruction and teaching, alignment between the district required processes and any new STEM initiatives would remove additional obstacles for teachers to implement the provided professional development. Expectations related to the teacher evaluation system should be responsive to teachers implementing new curriculum and instructional strategies (i.e. teachers should not feel that they might be penalized for not executing a "perfect lesson" when being observed, and any required templates or formats for submitting lesson plans should fit the work of the STEM professional development). Specific to STEM teaching and learning, teachers were afraid that more inquiry-based teaching might be viewed as "more chaotic" or "not following the expected, 'I do, we do, you do'" teaching model of the district. Finally, district-wide common assessments could be another factor that could present an obstacle for teachers to implement new STEM professional development ideas, particularly if they feel evaluative or do not align, either in terms of scope and sequence, or expectations for student outcomes.

Integrating STEM Curricular Resources

The district in this case study provides a number of curricular resources for teachers for STEM education through the district's website and intranet site for teachers. According to the district, the approach to science should be "inquiry-based" and "student-centered," and should be aimed towards supporting student achievement of the state content standards, which reflect the Next Generation Science Standards (NGSS) (NGSS Lead States, 2013). The resources include a curriculum guide for each grade level in K-8, which provides suggested time-frames for each unit, performance objectives, key terms and essential questions. In addition, lists of websites and other curricular supports are included for each unit.

It is interesting to note that even though the district provides a curriculum guide for each grade level, a majority of the teachers in the school did not know that this resource existed prior to being introduced to them in the professional development sessions, even though they had been teaching in the district for a number of years. When asked about the existing science curriculum in the district, the initial teacher needs assessment survey included responses such as "FOSS kits" (Full Option Science Systems, a kit-based science curriculum, developed by Lawrence Hall of Science, University of California, Berkeley, see www.fossweb.com) "we don't have a curriculum," or "we don't have any textbooks," with no teachers mentioning the district's curriculum guides.

For the purposes of the professional development in this study, the researcher and teachers at the school began with the district scope and sequence and curriculum guides, and then attempted to design sequences of instruction that incorporated language arts and mathematics standards in an inquiry-based approach, utilizing the 5E Instructional Model (Bybee, et al., 2006). As teachers were working with the curriculum guides, they found they still had a significant amount of planning to do before they could intentionally implement lessons in the classroom.

The large time-frames that were suggested for each unit in the districts' curriculum guide, often 10-12 weeks, meant that teachers needed to consider how to break up the time into specific activities that would support achievement of the standards and how to intentionally sequence and implement the activities using the 5E Instructional Model. This was a much more intensive planning process for most teachers compared to other subject areas where specific lessons were already laid out in much more detail, such as, for example, the mathematics curriculum, where teachers could look in the teachers' guide for suggestions for each day's lesson and they could go from lesson 5.1 to lesson 5.2 on the next day.

For example, one objective in the curriculum guide was that "students would be able to identify types of energy such as light, heat, sound and mechanical energy in order to explain how energy can exist in many forms." A teacher responded to that objective by finding an online worksheet that explained the different types of energy and using that worksheet as the lesson plan for that week's science lessons. Therefore, much of the professional development work in developing the curriculum was centered around how to sequence activities to support inquiry-based instruction. However, teachers noted during the grade level professional development time that this was a much different level of planning than they were used to planning in other subject areas.

In addition, the large number of performance objectives and essential questions that were provided for each unit tended to be overwhelming, and teachers struggled to pull everything into a cohesive sequence. During the professional development time, teachers also found it took a significant amount of time to vet the lengthy list of resources, and that a number of the provided links and suggested curricular resources were not available, outdated, or of questionable quality, which also led to frustration with the district provided resources.

The challenges relating to integrating and utilizing STEM curricular resources reflect the importance of providing high-quality curriculum resources and then appropriately communicating the curriculum and how to utilize the resources to all teachers. Providing teachers with a high-quality curriculum, much like the investments made for language arts and mathematics curricula, is critical so that teachers can spend more time developing their teaching pedagogy, rather than writing curriculum and investigating resources.

Managing the Competing Initiatives

Due to the complexities of a large district, often there are a number of different initiatives that are occurring at the same time, particularly because there is such a great need to improve student achievement all around. However, due to limited time and resources, teachers can only focus a certain amount of energy on each initiative each year.

In the district in this case study, the STEM school redesign initiative was launched in the same academic year as when a new mathematics curriculum and a new language arts series was introduced. As a result, teachers were being asked to become familiar with new curriculum in three different areas at the same time, limiting the amount of time they could focus on the STEM curriculum. An example of a specific challenge around these competing initiatives was that the teachers were required to attend a mathematics curriculum training and a language arts curriculum teaching on the two scheduled professional development days prior to the beginning of the school year. This left no time at the beginning of the school year to address STEM professional development, and the whole staff did not convene around the new STEM initiative until November of that school year. Common planning times for PLCs were divided between the different curricula, as well as other needed conversations around students and other school related business, leaving very little time to focus on developing STEM instruction.

At the district level, it is important to recognize that there is limited time to address different initiatives, and although they may all individually be important, it may be impossible to adequately focus on all of them at the same time. It is recommended that at the district level, there is oversight and conversation around prioritizing different initiatives, so that each can be given the appropriate attention to create quality and sustained change in teaching and learning.

IMPLICATIONS AND FUTURE RESEARCH DIRECTIONS

This chapter outlined a number of challenges to implementing STEM professional development, using an ecological systems perspective to frame the different levels of challenges that future professional development providers, and schools and district leaders who are considering implementing instructional and curricular change related to STEM education may face in urban schools. Within the chapter, specific suggestions and recommendations were made to address individual challenges.

However, the ecological systems perspective reminds researchers and practitioners that challenges come at many levels, and that all levels need to be considered for successful change to occur. It is not enough to focus only on the classroom level, or just the school or district level, because successful change needs to address concerns at multiple levels. Merely focusing on individual teachers is not sufficient if the teachers cannot effectively work within the school system. Similarly, it would not be wise to provide even an excellent district level curriculum without considering the structures within individual schools that would support implementation.

Within these different levels, there are also overlaps and intersections that can be leveraged. For example, the principal can help develop school-wide systems of leadership that engage teacher leaders and provide teachers with a sense of agency to instigate curricular change; this could impact both the classroom level and the school level. The principal can also mediate how district processes and initiatives are translated at the school level, through clear direction and leadership around the school's mission and vision.

These intersections suggest that the role of the building principal is critical to the success of school-wide initiatives. If professional development is to be successful, such as the professional development in this study which was created to support the redesign of the school into a STEM focused school, the school leader can be an important factor in alleviating many of the challenges that are present in urban school. Future research can focus more specifically on the role of the principal as a mediator of issues across multiple levels in the ecological systems model and how the principal can develop shared leadership that minimize some of these challenges.

This findings in this study were limited by several factors, and recommendations and implications should be applied with caution, with specific attention to the context of the site. Because this is a case study of one school, caution should be taken to avoid overgeneralization. Site specific considerations include the existing culture and climate of the school, structures and systems within the school, as well as individual personalities and interactions. Another limitation was that the professional development times and observation times were constrained by the school schedule. An additional limitation includes the potential for researcher bias, because the researcher also delivered the professional development. Future studies could include a wider number and range of schools, and/or a longer longitudinal study to examine the longer-term effects of STEM professional development.

CONCLUSION

This chapter provided insights to common barriers to conducting STEM professional development in an urban setting utilizing an ecological systems perspective. The chapter illustrated the complexity of integrating changes to STEM teaching practices along themes related to the classroom level, the school level, and the district level. In addition to addressing the specific challenges at each level, such as lack of teacher agency, limited resources, staff turnover, difficulties in scheduling, and competing initiatives, the ecological systems perspective reminds researchers and practitioners that challenges come at many levels, and that all levels need to be considered for successful change to occur. Leveraging the intersections between the different levels of classroom, school, and district can be one strategy to address the challenges in implementing instructional and curricular change related to STEM education.

ACKNOWLEDGMENT

Thank you to the school administration and teachers who so generously shared their time and welcomed me into their school and their classrooms.

This research received no specific grant from any funding agency in the public, commercial, or not-for-profit sectors.

REFERENCES

Abd-El-Khalick, F., Boujaoude, S., Duschl, R., Lederman, N. G., Mamlok-Naaman, R., Hofstein, A., ... Tuan, H. L. (2004). Inquiry in science education: International perspectives. *Science Education*, *88*(3), 397–419. doi:10.1002ce.10118

Anderson, R. D. (2002). Reforming science teaching: What research says about inquiry. *Journal of Science Teacher Education*, *13*(1), 1–12. doi:10.1023/A:1015171124982

Banilower, E. R., Heck, D. J., & Weiss, I. R. (2007). Can professional development make the vision of the standards a reality? The impact of the national science foundation's local systemic change through teacher enhancement initiative. *Journal of Research in Science Teaching*, *44*(3), 375–395. doi:10.1002/tea.20145

Borko, H. (2004). Professional development and teacher learning: Mapping the terrain. *Educational Researcher*, *33*(8), 3–15. doi:10.3102/0013189X033008003

Bragow, D., Gragow, K. A., & Smith, E. (1995). Back to the future: Toward curriculum integration. *Middle School Journal*, *27*(2), 39–46. doi:10.1080/00940771.1995.11496152

Bray, J. N. (2000). *Collaborative inquiry in practice: Action, reflection, and meaning making*. Thousand Oaks, CA: Sage Publications, Inc.

Bray, J. N. (2002). Uniting teacher learning: Collaborative inquiry for professional development. *New Directions for Adult and Continuing Education*, *2002*(94), 83–92. doi:10.1002/ace.62

Bybee, R. W., Taylor, J. A., Gardner, A., Van Scotter, P., Carlson Powell, J., Westbrook, A., & Landes, N. (2006). *BSCS 5E instructional model: Origins and effectiveness. A report prepared for the Office of Science Education, National Institutes of Health*. Colorado Springs, CO: BSCS.

Calvert, L. (2016). *Moving from compliance to agency: What teachers need to make professional learning work*. Oxford, OH: Learning Forward & NCTAF.

Capps, D. K., & Crawford, B. A. (2013). Inquiry-based instruction and teaching about nature of science: Are they happening? *Journal of Science Teacher Education*, *24*(3), 497–526. doi:10.100710972-012-9314-z

Cochran-Smith, M., & Lytle, S. (1999). Relationships of knowledge and practice: Teacher learning in communities. *Review of Research in Education*, *24*, 249–305.

Danter, E. H. (2005). *The intention-behavior gap: To what degree does Fishbein's integrated model of behavioral prediction predict whether teacher's implement material learned in a professional development workshop?* (Ed.D. Dissertation). Ohio State University. Retrieved from http://etd.ohiolink.edu/view.cgi?acc_num=osu1111698037

Darling-Hammond, L., & McLaughlin, M. W. (1995). Policies that support professional development in an era of reform. *Phi Delta Kappan*, *76*(8), 597–604.

Desimone, L. M., & Garet, M. S. (2015). Best practices in teachers' professional development in the United States. *Psychology. Social Education*, *7*(3), 252–263.

Desimone, L. M., Porter, A. C., Garet, M. S., Yoon, K. S., & Birman, B. F. (2002). Effects of professional development on teachers' instruction: Results from a three-year longitudinal study. *Educational Evaluation and Policy Analysis*, *24*(2), 81–112. doi:10.3102/01623737024002081

DuFour, R., & Eaker, R. (1998). *Professional learning communities at work: Best practices for enhancing student achievement*. Bloomington, IN: National Educational Service.

Ellis, A., & Fouts, J. (2001). Interdisciplinary curriculum: The research base: The decision to approach music curriculum from an interdisciplinary perspective should include a consideration of all the possible benefits and drawbacks. *Music Educators Journal*, *87*(22), 22–26, 68. doi:10.2307/3399704

Furner, J., & Kumar, D. (2007). The mathematics and science integration argument: A stand for teacher education. *Eurasia Journal of Mathematics. Science & Technology*, *3*(3), 185–189.

Garet, M. S., Porter, A. C., Desimone, L., Birman, B. F., & Yoon, K. S. (2001). What makes professional development effective? results from a national sample or teachers. *American Educational Research Journal*, *38*(4), 915–945. doi:10.3102/00028312038004915

Grossman, P., Wineburg, S., & Woolworth, S. (2001). Toward a theory of teacher community. *Teachers College Record*, *103*(6), 942–1012. doi:10.1111/0161-4681.00140

Gutherie, J. T., Wigfield, A., & VonSecker, C. (2000). Effects of integrated instruction on motivation and strategy use in reading. *Journal of Educational Psychology*, *92*(2), 331–341. doi:10.1037/0022-0663.92.2.331

Hurley, M. (2001). Reviewing integrated science and mathematics: The search for evidence and definitions from new perspectives. *School Science and Mathematics, 101*, 259–268. doi: .2001.tb18028.x doi:10.1111/j.1949-8594

King, K., & Wiseman, D. (2001). Comparing science efficacy beliefs of elementary education majors in integrated and non-integrated teacher education coursework. *Journal of Science Teacher Education, 12*(2), 143–153. doi:10.1023/A:1016681823643

Learning Forward. (2012). *Standards for professional learning*. Retrieved from: www.learningforward.org/standards/index.cfm

McHenry, N., & Borger, L. (2013). How can teacher-directed professional development lead to the identification, utilization, reflection on, and revision of 5E Learning Progressions? *Electronic Journal of Science Education, 17*(1), 1–24.

Minner, D. D., Levy, A. J., & Century, J. (2010). Inquiry-based science instruction—What is it and does it matter? Results from a research synthesis years 1984 to 2002. *Journal of Research in Science Teaching, 47*(4), 474–496. doi:10.1002/tea.20347

National Policy Board for Educational Administration. (2015). *Professional Standards for Educational Leaders 2015*. Reston, VA: Author.

National Research Council. (2000). *Inquiry and the national science education standards: A guide for teaching and learning*. National Academies Press.

NGSS Lead States. (2013). *Next Generation Science Standards: For States, By States*. Washington, DC: The National Academies Press.

Parson, E. C., & Summers, G. (2004). Use of images as reflective discrepant events: Pathways for elementary teachers to reconsider practice in relation to their views of science teaching and learning. *Electronic Journal of Science Education, 9*(1). Retrieved from http://ejse.southwestern.edu/article/view/7724/5491

Penuel, W. R., Gallagher, L. P., & Moorthy, S. (2011). Preparing teachers to design sequences of instruction in earth systems science: A comparison of three professional development programs. *American Educational Research Journal, 48*(4), 996–1025. doi:10.3102/0002831211410864

Petrides, L. A., & Guiney, S. Z. (2002). Knowledge management for school leaders: An ecological framework for thinking schools. *Teachers College Record, 104*(8), 1702–1717. doi:10.1111/1467-9620.00217

Priestley, M., Edwards, R., Priestley, A., & Miller, K. (2012). Teacher agency in curriculum making: Agents of change and spaces for manoeuvre. *Curriculum Inquiry*, *42*(2), 191–214. doi:10.1111/j.1467-873X.2012.00588.x

Roth, K. J., Garnier, H. E., Chen, C., Lemmens, M., Schwille, K., & Wickler, N. I. (2011). Videobased lesson analysis: Effective science PD for teacher and student learning. *Journal of Research in Science Teaching*, *48*(2), 117–148. doi:10.1002/tea.20408

Smith, J., & Karr-Kidwell, P. (2000). *The interdisciplinary curriculum: A literary review and a manual for administrators and teachers*. Retrieved from ERIC database. (ED443172)

Stoll, L., Bolam, R., McMahon, A., Wallace, M., & Thomas, S. (2006). Professional learning communities: A review of the literature. *Journal of Educational Change*, *7*(4), 221–258. doi:10.100710833-006-0001-8

Supovitz, J. A., & Turner, H. M. (2000). The effects of professional development on science teaching practices and classroom culture. *Journal of Research in Science Teaching*, *37*(9), 963–980. doi:10.1002/1098-2736(200011)37:9<963::AID-TEA6>3.0.CO;2-0

Thompson, S. C., Gregg, L., & Niska, J. M. (2004). Professional learning communities, leadership, and student learning. *RMLE Online: Research in Middle Level Education*, *28*(1), 15. doi:10.1080/19404476.2004.11658173

Van Driel, J. H., & Berry, A. (2012). Teacher professional development focusing on pedagogical content knowledge. *Educational Researcher*, *41*(1), 26–28. doi:10.3102/0013189X11431010

Weinbaum, A., Allen, D., Blythe, T., Simon, K., Seidel, S., & Rubin, C. (2004). *Teaching as inquiry*. New York, NY: Teachers College Press.

Yin, R. K. (2009). *Case study research: Design and methods* (4th ed.). Thousand Oaks, CA: Sage Publications.

Zepeda, S. J., Parylo, O., & Bengtson, E. (2014). Analyzing principal professional development practices through the lens of adult learning theory. *Professional Development in Education*, *40*(2), 295–315. doi:10.1080/19415257.2013.821667

Zhao, Y., & Frank, K. A. (2003). Factors affecting technology uses in schools: An ecological perspective. *American Educational Research Journal*, *40*(4), 807–840. doi:10.3102/00028312040004807

KEY TERMS AND DEFINITIONS

District Initiatives: Plans or strategies implemented at the district level that impact teaching and learning within schools.

Integrated Curriculum: A curriculum that incorporates literacy and math standards within the STEM curriculum.

Principal Leadership: Actions of the building leader that provide guidance and direction to improve teaching and learning within a school.

Professional Development: Opportunities for teachers and staff to engage in learning specific to topics related to improving their curriculum and delivery of instruction.

Professional Learning Communities: Groups of teachers working together to improve teaching and learning within their classrooms.

STEM Curriculum: Curriculum integrating science, technology, engineering, and math teaching and learning standards.

Teacher Agency: A teacher's willingness and ability to take action and make changes.

Teacher Leadership: Ability of teachers to influence teaching and learning beyond their own classrooms.

Urban Education: Schools in large, densely populated areas with diverse populations, typically characterized by large enrollments and complex bureaucratic systems.

Chapter 4
The Use of Instructional Coaching and Analogy to Enhance STREAM Professional Development for Teacher Quality Improvement

Otto Carl Wilson Jr.
Catholic University of America, USA

Seidah Armstrong
Performance Management Services LLC, USA

ABSTRACT

The design, implementation, and continuous improvement of professional development (PD) is a key component to address inherent challenges in helping our urban schools achieve their full potential for STREAM (STEM with integrated reading and arts) learning. While there are many types of PD, instructional coaching is one of the most promising strategies that can address both teacher and student needs on multiple levels. The DC FUSION STREAM PD Consortium was formed in 2017 with initial seed funding from the US Department of Education's Teacher Quality Improvement Grant Program. In this chapter, key aspects of the DC FUSION STREAM PD Consortium are described in the context of analogies for enhanced STREAM PD and how a broader view and implementation of instructional coaching can be used to transform teacher practice and effectiveness and elevate student achievement.

DOI: 10.4018/978-1-5225-7814-7.ch004

INTRODUCTION

Washington, DC is a rapidly expanding city that hosts many large-scale construction and renovation projects that are in various stages of completion. While these visible changes help to improve the physical infrastructure of DC, there are several important efforts that focus on improving the very important educational and learning infrastructures in DC. The similarities between construction, renovation projects, and education reform can be used to build analogies and provide insight to address the challenges that are encountered in K-12 STREAM Education in DC where STREAM refers to STEM enriched with reading and arts. Analogies are powerful learning tools (Pollack, 2015; Vendetti et al., 2015) that can help foster a deeper understanding of complex phenomena by linking simple concepts and images together in creative ways to help learners make sense of abstract concepts. The human brain is hardwired to learn from experience. As such, the mental models that develop with analogies can empower learners to more effectively take simple concepts and assemble them into elegant ideas to solve real world problems (National Research Council, 2000; Matlen et al., 2011; Nature Photonics Editorial, 2014). Boaler (2016) discusses brain research as it relates to comprehending mathematics. Moreover,this building block approach that combines brain research and analogies can help teachers design engaging experiences in STREAM learning environments.

School reform efforts often include teacher professional development (PD) as a means for improving schools and impacting student achievement. Professional development is a key framework component for achieving goals related to improved teacher practice and student learning outcomes. There are a variety of ways to present professional development. Some of the more common PD approaches include content specific workshops, full-day professional development sessions, peer mentoring, coaching, and blended professional development that consists of independent workshops and coaching or peer-mentoring. Joyce and Showers (2002) identified four training concepts that are very important components in instructional coaching: Knowledge, Modeling, Practice of the Skill Set, and Peer Coaching. Inclusion of these concepts within professional development interventions allows educators to engage in the presented content in a more meaningful way. The authors suggest that teachers should engage in practicing the newly learned skill, concept or strategy for 8-10 weeks or for about 25 trials to improve retention (Joyce and Showers, 2002). As mentioned in the above construction analogy, there must be an organized structural approach to implementing professional development that will yield results that transform classrooms. Educators must be proactive in conducting careful analyses of current educational ecosystems in order to more carefully identify the best practices for the design, implementation and assessment/evaluation of teacher PD and the conditions under which PD has the greatest impact on student learning and achievement.

The purpose of this chapter is to describe how a *Teacher Quality Improvement* (TQI) Grant provided funds for a group of university faculty and STREAM professionals in Washington, DC to form a consortium to initiate a collaborative design process to provide high quality STEM based professional development for teachers in a large urban education environment. Fusion energy and the inspiration which it generates served as the focal point analogy and theme for designing the *DC Focused Universe of STREAM Inspiration with Optimized Nurture (FUSION) STREAM Professional Development (PD) Consortium*. Instructional coaching was featured as a central component to support participating teachers in their journey towards implementing key STREAM concepts that were introduced during a series of professional development workshops conducted from August 2017 through June 2018. The DC FUSION STREAM PD Consortium received one of the last TQI Grants awarded in Washington, DC in 2017 under the *No Child Left Behind Educational Act* of 2001.

This chapter focuses on distilling the main points of the DC FUSION STREAM PD Consortium and is organized in the following manner. In the first section, a brief literature survey regarding instructional coaching and its prominent role in professional development efforts to help bridge the academic achievement gap is presented. Next, aspects of multi-year, large scale PD efforts entailed in the *Measures for Effective Teaching* (MET) Project, a model project, are highlighted and how some of the conclusions drawn from the MET Project support the need for a more active implementation of instructional coaching in comprehensive PD efforts (Will, 2018). The final section focuses on the initial development of the DC FUSION STREAM PD Consortium and highlights the work and achievements during the first year. While the DC FUSION STREAM PD Consortium faced its share of challenges, lessons related to the importance of clear communication, vision, and teamwork were reinforced throughout the initial year and will be valuable for achieving the ambitious goal of providing high quality STREAM PD for all metropolitian District of Columbia teachers.

BACKGROUND

Achieving the Goal of All Students Reaching the Top

In 2010, President Obama addressed America's global economy programs and clearly identified STEM education as a part of the needed solution. President Obama's initiative to catapult America's students from the center to the head of the pack involved improving and increasing STEM literacy, engaging underrepresented groups in STEM education fields, and creating more STEM career opportunities for Americans (President's Council of Advisors on Science and Technology, 2010; National Research Council, 2011; and Zuger, 2012). Identifying underrepresented groups is an important step in addressing critical issues in educational settings. One prevalent issue is the academic achievement gap among minority students (Gordon, 2001).

The academic achievement gap is persistently manifested as lower standardized test scores for students of color (Williams, 1996). Due to the disproportionately large percentage of historically underrepresented students in urban schools, the development of strategies to bridge the academic achievement gap and help all students learn to their fullest potential is critically important to improve urban STEM education (Wang and Kovach, 1995). Research on the achievement gap regarding finding strategies to bridge the gap by Gordon et al (2001; Bennet et al, 2004) have yielded important models such as the Affirmative Development of Academic Ability and Intellectual Competence Model (Bennett et al., *2004).* This work provides a very powerful model to address inherent challenges that contribute to achievement gaps. The specific capabilities that are exhibited by students with academic promise, according to the work, include 1) Literacy and numeracy, 2) Mathematical and verbal reasoning, 3) Skill in creating, recognizing, and resolving relationships, 4) Problem solving from both abstract and concrete situations, as in deductive and inductive reasoning, 5) Sensitivity to multiple contexts and perspectives, 6) Skill in accessing and managing disparate bodies and chunks of information, 7) Resource recognition and utilization (help seeking), and 8) Self-regulation (Bennett et al., 2004). When addressing critical school reform needs such as closing achievement gaps in American schools, one should consider STEM content to creatively expose students to real world problems to spark interest through inquiry and problem-solving.

Many issues that contribute to academic achievement and associated gaps are related to the dynamics of teacher environment and student interactions. Efforts to improve these dynamics and enhance the learning environment can be facilitated by improving the quality of teacher interactions with students through enhanced professional development and its potential for improving student achievement (Darling-Hammond and McLaughlin, 1995; Joyce and Showers, 2002). An important question that needs to be answered to achieve goals in enhancing urban STEM education is "How do we fully assess the true impact of professional development interventions on teacher efficacy and student learning outcomes?"

Professional Development or Pedagogical Discontent

In a 2012 study, the i-STEM Summer Institute was established to introduce STEM concepts to teachers in grades 4th-9th (Nadelson et al., 2012). Some goals of the session were to develop a higher level of teacher capacity for STEM instruction and to gather empirical data to establish its effectiveness (Nadelson et al., 2012). There is evidence that support teacher's effectiveness and their emotional connections towards the content (Darling-Hammond and Bransford, 2007). The i-STEM Institute sought to break down barriers by equipping teachers with knowledge and tools to make them confident in their classroom delivery of STEM concepts. By doing this, the PD would dissolve the "pedagogical discontentment" that exists when educators experience discomfort due to lack of competence in a subject matter (Nadelson et al., 2012). Many teachers believe professional development sessions are not designed for their specific content. Pedagogical discontentment occurs when teachers do not connect to content. As a result, many educators return to classrooms after participating in professional development events altering classroom practices slightly if at all.

Similar research has also utilized modeling and practice to support teachers in professional development. In 2012, 51 secondary science educators partook in a gaming design-based research study that utilized the Technological Pedagogical Content Knowledge (TPACK) Framework to gauge the level of content knowledge, pedagogy, and technology integration, as teachers designed a science-based video game. The research focused on teacher efficacy surrounding five factors: 1. General Computer Use, 2. Science Learning, 3. Inquiry Teaching and Learning, 4. Synchronous chat/text, and 5. Playing Video Games (Annetta et al., 2013). Annetta et al.'s (2013) study also revealed the pedagogical discontentment that is present when educators are uncomfortable with content. Results revealed that teachers who spent more time practicing and using technology daily had higher efficacy using an inquiry-based approach to teaching. Providing a hands-on structured approach in professional development led to teacher buy-in that resulted in classroom implementation of new knowledge.

The Measures for Effective Teaching (MET) Project

In 2009, the Bill and Melinda Gates Foundation funded a large-scale project called Measures for Effective Teaching (MET). The goal of the MET Project was to improve the quality of information related to teaching effectiveness that is available to education professionals within states and school districts. This valuable information would help in the construction of fairer and more reliable systems for measuring teacher effectiveness with enhanced features for feedback, development and continuous improvement (MET Project Research Paper, 2010).

As part of the MET Project, nearly 3000 volunteer teachers from six predominantly urban school districts in six states across the country helped to identify better approaches to teacher development and evaluation. The initial participating districts included Charlotte-Mecklenburg, NC Schools, Dallas, TX ndependent School District, Denver, CO Public Schools, Hillsborough County, FL Public Schools (including Tampa, Florida), Memphis City, TN Schools, and the New York City Department of Education. As part of the project, multiple data sources were collected and analyzed over three school years, including student achievement gains on state assessments and supplemental assessments designed to assess higher-order conceptual understanding; classroom observations and teacher reflections on their practice; assessments of teachers' pedagogical content knowledge; student perceptions of the classroom instructional environment; and teachers' perceptions of working conditions and instructional support at their schools (MET Project Research Paper, 2010).

The three premises of the MET project were:

1. Whenever feasible, a teacher's evaluation should include his or her students' achievement gains.
2. Any additional components of the evaluation (e.g., classroom observations, student feedback) should be demonstrably related to student achievement gains.
3. The measure should include feedback on specific aspects of a teacher's practice to support teacher growth and development (Met Research Paper, 2010).

Research has shown that ineffective teacher behaviors must be addressed through intentional and supported professional development in order to improve teacher effectiveness and increase student achievement, (Boaler, 2016; Joyce and Showers, 2002). The MET Project provided guidelines for the DC FUSION STREAM PD Consortium Project that highlight the important role that instructional coaching can play in addressing some of the weaknesses that were pointed out in recently released program evaluation reports (Will, 2018). With the ultimate goal of meeting the STREAM professional development needs for all teachers in the DC metropolitan area, the DC FUSION Project can serve as a tool to design and cultivate a professional development model infused with instructional coaching that will help transform classrooms through improved teacher effectiveness and increased student enagagement and achievement.

DC STREAM FUSION PD CONSORTIUM AND THE TQI GRANT: THE NAME AND PURPOSE

Fusion is a process in which small atoms combine to form larger atoms and release large amounts of energy. It is the opposite of fission reactions in which large atoms such as Uranium (U) are broken apart into smaller atom-based reaction products. Fusion processes provide power to light up stars. Each point of light in the night sky is a star that is located extremely far from Earth, yet we can see them and ask questions about their origins. Scientists have reported evidence for observing the first light that illuminated the Universe in the early dawn of creation. This initial light has been attributed to the fusion reactions that gave birth to the first stars and the full story of this phenomenal discovery includes theorized contributions from dark matter (Bowman, Rogers, Monsalve, Mozdzen, & Mahesh, 2018; Barkana, 2018). All stars that we see undergo fusion reactions and these processes provide the driving force for change and developments in our Universe ranging from subatomic particle phenomena involving quarks and neutrinos to gigantic quasars which can exist at the center of massive black holes. Most galaxies host massive black holes in their centers which play an important function in recycling matter and destroying as well as giving birth to stars.

Fusion energy and all the inspiration which it generates serves as the focal point and theme for a Teacher Quality Improvement (TQI) grant that was awarded to a consortium of DC universities and STEM professional organizations to address challenges in STEM education--the DC FUSION STREAM PD Consortium. The initial award was granted to the Catholic University of America in partnership with George Washington University, Trinity University, Howard University, and educational consultants: Performance Management Services, LLC, in 2017. The initial cohort of schools that were recruited to participate in the pilot TQI professional development workshops included the Preparatory School of DC, Calvary Christian Academy, and Cardozo High School. Due to changes in the level of school participation, Sacred Heart was added midway through the initial year. As a result of the additional school, the TQI grant hosted two cohorts of teacher participants. This was done to ensure that the newly adopted school received the same professional development sessions as were conducted with the first cohort of teacher participants.

The goal and mission of the DC FUSION STREAM PD Consortium is to provide access to the best STREAM professional development opportunities and learning resources for teachers and students in the DC metropolitan region—the TQI grant fulfilled part of this overall mission. The DC FUSION vision involves seeing all teachers and students learning at their full potential and fully engaged in exploring all aspects of our expanding Universe (from quarks to quasars) based on current and future iterations of high-quality science learning standards, such as the Next Generation Science Standards (NGSS) (National Research Council, 2012, 2013) and long term science literacy projects such as the American Association for the Advancement of Science (AAAS) Project 2061 (AAAS, 1989, 1993, 2001a, 2001b; Koppal and Caldwell, 2004). The objectives of DC STREAM FUSION Consortium are: to 1) Form a Consortium of DC Universities, K-12 Schools, and STEM Professionals which can form a "Critical Mass" to develop into a self-sustainable STREAM PD Wave that will provide High Quality STREAM PD for all schools in DC, 2) Develop a STREAM Question Tree that integrates Past, Present, and Future STREAM Questions into a framework that will promote finding solutions to the great challenges which face our learning community, 3) Develop a FUSION Assessment Matrix to evaluate the current status of STREAM Learning in DC Schools and the opportunities that exist for enhanced learning,and 4) Create dynamic learning strategies and plans to move schools along the pathway from stellar dust clouds to fully functioning stars which radiate enormous levels Unlimited inspiration and achievement.

Funding and Structure

The initial TQI grant provided seed funding in the amount of approximately $175,000 to help build the DC STREAM FUSION Consortium and to take a critical look at determining the conditions which lead to forming the "Critical Mass" which leads to achieving full scale fusion. There have been many efforts to develop large scale collaborative projects in many different fields of endeavor, but most fail to achieve the elusive status of fusion. A key part of our project approach involves developing a "question tree" that serves as a framework for investigating the multi-faceted challenges which influence our ability to most effectively address teacher PD needs in a way that will result in concrete student achievement gains that will bridge academic achievement gaps. The roots of the question tree represent past questions and how they inform our ideas, the trunk and branches represent present questions and the opportunities that they provide, and the leaves and fruit represent future questions and the possibilities and potential which exist through channeled inspiration and creativity. A tree is prominently displayed in a memorial to Dr. George Washingtion Carver to commemorate the 150th anniversary of his stay in Winterset, Iowa (Figure 1).

Figure 1. Ceramic Tile Mural created by school children to commemorate the 150th anniversary of Dr. George Washington Carver and his contribution to the community of Winterset, Iowa. (Photo Credit: Debra Wilson, 2018)

The DC STREAM FUSION Consortium's response to the grant ensured that PD would incorporate coaching, modeling, and administrative coaching for STEM content sustainability. The participant teacher time commitment included the following:

- 15 Hours After School Professional Development Commitment
- 10 Hours of Online Project Commitment
- 30 Hours Full-Day STREAM Professional Development Boot Camp
- 5 Coaching Visits to Classroom w/ Post-Reflections and Lesson Planning Support

The first goal of the project was to provide free quality professional development by exposing educators to STREAM strategies in the classroom to help close the achievement gap. The second goal was to change teacher behavior in the classroom towards STREAM concepts by providing strategies that promoted engagement through more student-centered activities. The third goal was to provide engaging professional development sessions, in which teacher participants were immersed in sample lessons and reflected on their experiences as students. During this time teachers reflected on how similar lessons could be used to enhance their own classrooms and to promote student mastery of key STREAM concepts. Each building was also assigned an instructional coach. The role of the instructional coach was to provide guidance, support, and motivation to DC STREAM FUSION Consortium's participants in their quest to implement STREAM concepts and strategies introduced during PD sessions. In addition, a leadership coach provided general guidance to school leaders so that leaders would be able to support DC STREAM FUSION Consortium's participants and identify key STREAM strategies and concepts when visiting classrooms. It is important to note that building leaders did not participate in the professional development portion of the project. However, it was imperative that building leaders received general knowledge of the project's intended outcome.

The TQI grant worked with an external evaluator to assess the project.

The external evaluator gathered preliminary data through a survey administered to participants via email or at the first PD session in August, 2017. The same survey was administered again at the conclusion of the project in June, 2018. Over the course of the year, the evaluator visited professional development sessions and interviewed participants. The evaluator also documented verbal testimonies of teachers, school leaders, and project facilitators to gauge the success of the project upon completion.

The professional development workshop sessions were structured in the following phases:

Phase I (Nebula): Introduction to STREAM Symposium Workshop (August, 2017)

- Introduction to STREAM, Engineering Design Process, 4C's, STEM Careers, and standards ion the classroom

Phase II (Turbulence/Condensation): After School PD Workshops (Fall, 2017 through Spring, 2018)

- STEM Across the Content Areas
- Creating a STREAM Culture in School
- STREAM and the Community: Addressing Colleagues, Students, Parents, Building Leaders, and Community Partners and hosting School STEM Nights
- STEM Stations in the Classrooms and the Role of Technology in Student Learning
- Science Fairs vs. STREAM Fairs: Revamping Interests and the Benefits for America's Youth

Phase III (Protostar): Online Lesson Planning PD Workshops (Fall, 2017 Through Spring, 2018)

- Infusing Technology into Your Lessons: QR Codes, Digital Portfolios and More
- Building a Short Circuit w/ Foil
- SeeSaw Reflections on Classroom Transformation through STEM Related Articles
- Creating Lesson Plans Using the 4C's (Collaboration, Creativity, Communication, and Critical Thinking)
- Creating an Energy Source

Phase IV (Fusion): DC FUSION STREAM PD Boot Camp (Summer 2018)

- Addressing the M in STEM
- Earth Science and Renewable Resources: Project Based Learning (PBL) & STEM
- The Engineering Design Process and Rube Goldberg
- The Scientific Argument Driven Inquiry Instructional Model
- STEM & Science Strategies, Resources and Activities

The DC STREAM FUSION Consortium developed a series of professional development workshops that encompassed the initial introduction to the DC STREAM FUSION Consortium workshops--10 monthly workshops (both in person and online) that intentionally engaged participants in various aspects of STREAM and how it can be integrated across the curriculum, and the culminating DC FUSION STREAM PD Boot Camp Workshop. The professional development sessions led teachers through several Science, Engineering, and Math focused activities. Topics included creating electric circuits, building Rube Goldberg machines, constructing sustainable towers, geospatial landmark visits, Physics in the PreK-12 classroom, and home designs with math themes. Each of the workshops was developed and implemented to include the following four phases as described below.

Phase I of the DC FUSION STREAM PD program consisted of a week long introduction to STREAM Symposium workshop which started in August, 2017. Teachers were exposed to key elements of STEM such as the Engineering Design Process, the 4C's (Communication, Collaboration, Critical Thinking, and Creativity), STEM careers and real-world connections, and exploring the role of the standards-driven classroom during engaging workshops. Phase I provided an opportunity for the DC FUSION STREAM PD Consortium facilitators to collect formal and informal data regarding the various levels of STREAM exposure and experiences the participants possessed. Phase I can be thought of as the nebula coalescence phase that occurs during star formation. In this stage, the initial gas cloud forms and begins to interact and condense.

Phase II consisted of five after school professional development workshop sessions. These five sessions were designed to provide more in-depth knowledge on how to use strategies and key concepts to construct engaging lessons to promote academic success and content mastery in the classroom. The sessions were spread out over the course of the 2017-18 academic year and are shown in Table II. Phase II can be viewed by analogy as the turbulence stage, when structures start to form within the nebula dust cloud and portions of the dust region become heavy enough to collapse under their own mass.

During Phase III, FUSION teachers participated in a series of online lesson planning PD workshops. The purpose of the online professional development workshop was to extend the after school sessions to allow teacher participants the opportunity for hands on engagement with STREAM activities that included constructing circuits and creating workable energy sources. The online sessions also encouraged teachers to reflect on professional development experiences by reading related articles and documenting their experiences as student learner. The online sessions also provided a platform for teachers to share how their PD experience transformed lessons for their students. In the fusion analogy, Phase III represents the formation of a protostar which is the precursor to a star.

Phase IV consisted of the DC STREAM FUSION Consortium's Summer PD Boot Camp. This was the culmination of the teacher participation component of the DC STREAM FUSION Consortium. The week-long camp for teacher participants allowed the DC FUSION STREAM PD Consortium to provide an immersive experience for cohort participants and some of their colleagues from their school who showed interest in learning more about STREAM concepts and how to infuse STREAM strategies into their student lessons. Phase IV can be thought of as the initiation of fusion in which DC STREAM FUSION Consortium teachers have been cultivated and empowered to self ignite and form a shining star of STREAM inspiration for enhanced classroom learning.

Following each professional development session, teacher participants reflected on how the activity or similar STEM based activities could be used to create student-centered opportunities in their classroom to help support content mastery. Identifying applicable standards was key in this process. DC STREAM FUSION Consortium participants collaborated with colleagues to design standards-driven lessons. Teachers also provided essential questions for students to consider while participating in the lesson.

Another key STEM element that was an essential part of the workshops was to address links between the associated careers and the presented content. According to the National Science Foundation, Women occupied only 28% of the science and engineering workforce in 2010 (NSF, 2014). The number of Minorities entering STEM careers has also displayed low numbers in past years. Exposing youth to STEM careers early and often could increase the desire for students to choose STEM careers later in life. The DC STREAM FUSION Consortium workshops consistently showcased STEM based career opportunities that coincided with the content and activities being presented. Articles on the STEM workforce gap were shared with participants. DC STREAM FUSION Consortium participants were encouraged to create a timeline of careers that were introduced during the year and to post the STEM career timeline in the classroom. DC STREAM FUSION Consortium facilitators and coaches also encouraged teachers to find community representatives with a particular career to visit the classroom and meet with students or post the community representative's picture and job description in the classroom.

The DC STREAM FUSION Consortium used analogies over the course of the year to explicate the role of the urban teacher in the STREAM infused classroom. A few of the analogies included modeling each school in DC as a star in a large disctrict wide constellation. The intensity of each school is a reflection of the school's level of engagement with STREAM instruction. The level of fusion can also be viewed related to classrooms within the schools, the teachers within the classroom, and ultimately the students who are the most important beneficiaries of the DC STREAM FUSION Consortium. Under the right conditions, each student can become a brightly shining star of STREAM inspiration.

The Importance of Technology in STREAM Professional Development

Darling-Hammond and McLaughlin (1995) discussed ideas related to the importance of clearly defining the role of professional development and creating a "learner-centered professional development" environment and how to fully include teachers as active participants. In response to the call for active learner-centered professional development, the DC FUSION STREAM Consortium used key aspects from the "learner-centered professional development" model of Darling-Hammond and McLaughlin (1995) in the PD design process as listed below:

1. The role of teachers as learners and teachers
2. New designs for professional development
3. School culture and how it relates to critical inquiry
4. Elements of effective curriculum frameworks messages emerging for policy-makers and administrators concerned with professional development for teachers

An essential component of the DC STREAM FUSION Consortium focused on the utilization of technology. Understanding how to incorporate technology effectively meant providing a clear understanding of the purpose of technology in instruction. The DC STREAM FUSION Consortium developed workshops that intentionally engaged participants in working with digital tools, apps, and software that allowed them to create authentic work. The Substitution, Augmentation, Modification, and Redefinition (SAMR) Model, which focuses on how technology is used in the classroom, was introduced to DC FUSION Teachers to help support the technological growth of participants. The SAMR Model (Puentedura, 2014) was used as a framework to provide a guide for teachers on how to develop and design digital learning experiences for students and how to utilize technology to enhance learning. The definition of each stage in the SAMR model is shown below.

- **Substitution:** Technology is used as a direct substitute with no functional change
- **Augmentation:** Technology acts as the direct tool substitute with functional improvement
- **Modification:** Technology allows for significant task redesign
- **Redefinition:** Technology allows for the creation of new tasks.

The SAMR Model explicitly shows how technology can be used and the intended outcome that occurs when using technology as defined in each stage. While the concepts in the SAMR model are very engaging, it is important to note an important caveat in regards to the development of this model. There are some criticisms regarding the level of scholarly citations and pedagogy related to the SAMR model (Green, 2014).

DC FUSION Participants were provided an opportunity to share how they infuse technology into their lessons, prior to introducing the SAMR model. After a clear definition was established, teachers were engaged in another discussion. Many teachers admitted that most of their technology infused lessons were at the substitution stage. Even those participants familiar with the SAMR Model stated they rarely moved past the Augmentation Stage of SAMR. Participants were assigned several tasks requiring the use of technology to show work and mastery of key STREAM concepts.

Another conceptual model that the DC FUSION STREAM Consortium used to help our teachers walk away with a clearer understanding of key ideas was the 4C Model which incorporates Collaboration, Communication, Critical Thinking and Creativity as it relates to education. The concept of the 4C's in education derived from a 2002 study in which 18 essential skills were identified as key 21st Century learning skills in education (National Education Association, 2012). Some years later, 21st Century leaders would reconvene to highlight the most critical 21st Century learning skills for K-12 classrooms, honing in on the 4C's (National Education Association, p. 3). Using the inquiry-based learning approach, DC STREAM FUSION Consortium participants were provided with a research question within the 4C Model framework and created "paper slides" that highlighted their research. A "paper slide" is a video recorded presentation using words and images on several sheets of paper to represent each concept, while verbally explaining each illustration. Mobile devices were used to present their findings. Other examples included utilizing the video feature of Seesaw, which is an online digital portfolio with a student and parent friendly app to create, access, and discuss content. DC STREAM FUSION Consortium participants used SeeSaw to record their understanding of creating simple circuits and as a reflective tool throughout the project.

After DC STREAM FUSION Consortium teachers had the opportunity to clearly understand how students use technology to create authentic work, they were allowed to experience the SAMR model at each level as active participants. Each teacher created three lessons that demonstrated how students use technology at the Modification, and Redefinition Stage. Providing a firm foundation in STEM meant ensuring that all pieces of the SAMR Model were addressed while working with DC STREAM FUSION Consortium teachers to minimize gaps in knowledge due to lack of previous exposure.

Identifying Key Characteristics of STEM and Instructional Fusion

The introduction of key STEM characteristics was an imperative component for creating a firm foundation for STREAM Learning opportunities for participants. Teachers who have a clear understanding of what instructional practices entails and how it benefits the intended audience are more effective at helping their students make gains and close the achievement gap. DC STREAM FUSION Consortium participants were challenged to infuse concepts learned in their PD sessions into the classroom. One of the first elements of STREAM pedagogy that was discussed involved the importance of reviewing previous lessons to identify how the 21st Century learning skills creativity, collaboration, critical thinking, and communication (the 4C's) were addressed. DC STREAM FUSION Consortium participants openly admitted that previous efforts to design lessons did not include the 4C's. DC STREAM FUSION Consortium participants were given the opportunity to enhance previous lessons by incorporating the 4C's. DC STREAM FUSION Consortium instructional coaches reinforced this instructional practice during the observatory stage of the visit. Coaches took copious notes in order to 1. Identify areas in the lesson where students were engaged in one or more of the 4C's, and 2. Specify areas in the lessons with provided examples of how utilizing the 4C's could have enhanced students' learning experiences.

RESULTS AND DISCUSSION

The Important Role of Coaching in Professional Development

There are several characteristics of effective coaches that can be cultivated in STREAM teachers to enhance and improve teacher effectiveness: 1) Attention to Detail (Observation/Collaborative Reflection), 2) Deep Subject Matter Knowledge (Content Matter Expert or Research/Professional Learning Community Liaison),3) Ability to clearly communicate concepts (Model/Co-teach), 4) Reason, relevance, and practicality, 5) Great Motivational Skills, 6) Vision and foresight – being able to see potential that is clearly evident as well as hidden "diamond in the rough" talent, 7) Patience, and 8) Long Term Commitment.

A broader perspective of coaching as an effective tool to help all teachers reach their full potential needs to be embraced. Research continues to support instructional coaching as a professional development technique to improve teacher classroom practices, thus improving student learning (Kowal and Steiner, 2007; Grant, 2013; Kraft, Blazar, and Hogan, 2018). When schools establish authentic professional learning communities, where teachers have open dialogues about instructional best practices, a shift in classroom practices will take place (Ertmer et al., 2005). The National Staff Development Council document on standards for staff development states, "Staff development that improves the learning of all students requires resources to support adult learning and collaboration" (National Staff Development Council, 2001, p. 3).

Each DC FUSION Grant participating school was assigned one instructional coach. The instructional coach visited five times during the academic year. Each instructional coach spent the equivalent of one school day per visit supporting DC FUSION Participants, through observations, modeling, co-planning, and reflection. Prior to working at schools, each DC FUSION Instructional Coach received directives to emphasize the importance of establishing a 4C's poster, Engineering Design poster, and Scientific Method poster for school wide adoption. Coaches also encouraged building level leaders to lead the way with these instructional practices becoming schoolwide practices. DC FUSION Participants experienced hands-on opportunities with similar posters during evening professional development sessions. The session facilitator stressed the importance of providing a level of consistency for students as they moved through the school. A schoolwide adoption also promoted a common instructional language. DC FUSION Instructional Coaches reinforced the importance by having DC FUSION Participants consciously incorporate these posters into their lessons, so that students could have multiple opportunities to become familiar with the process, steps, and concepts that are illustrated in these important scientific and engineering practices.

In addition to focusing on the 4C's, coaches also emphasized the importance of developing STEM infused lessons and activities based on standards such as Common Core or NGSS. Coaches worked with DC FUSION Participants to address content standards and the role STREAM plays in supporting standards. DC FUSION Participants presented lessons during coaching sessions, with standards and essential questions already developed. At times DC FUSION Instructional Coaches realized that deconstructing standards was necessary to create engaging lessons for students with clear expectations for student learning. If lessons entailed students creating a physical final product, coaches encouraged DC FUSION Participants to adopt and utilize an Engineering Design Process model, which provides a step by step process of how to create a product. Likewise, if a lesson called for introducing the Scientific Method, coaches directed participants to include the steps as an integral part of the class.

It was important for DC STREAM FUSION PD Consortium to clearly explicate the desired intended outcome and expectations of the DC FUSION Instructional Coach. All assigned coaches came with prior coaching experiences. *The Art of Coaching: Effective Strategies for School Transformation* (Aguilar, 2013) and *Coaching Classroom Instruction* (Marzano et al., 2013) were selected as the primary frameworks used to guide DC FUSION Instructional Coaches. Each coach used a Padlet page to post artifacts from each visit. Artifacts included teacher lessons, photos, videos, and notes. DC FUSION Instructional Coaches emailed an agenda prior to the visit to the building level administrators and cohort participants. End of the day debriefing sessions with building level administrators were conducted by DC FUSION instructional coaches. This was done in coordination with submission of a post-reflection summary. Coaches used a STEM Observation form to document visits. Coaches also provided reflective thoughts at the end of each visits in an email to designated administrators.

An important component of the DC FUSION Grant process involved reflecting on the STEM workshops and experiences. Teachers worked fervently to transform their classrooms. Following the workshops teachers were provided homework, which often included an article reflection piece. Participants used Seesaw to journal their thoughts. Coaches would view the reflections on Seesaw and create topics of discussion based on the participant's thoughts during visits.

Coaches were instrumental in assisting participants in incorporating various strategies introduced during workshops into their classroom lessons. For example, all participants interacted with Seesaw, Padlet, and QR Codes during the evening workshops. Coaches utilized the workshop slides to highlight how each digital tool could be used to enhance student learning. Many participants took interest in Seesaw because it serves as a digital portfolio. The immediate impact of the introduction of digital tools and reinforcement of intended outcomes from the coach using the SAMR model, resulted in several teachers introducing the journal and video feature to students.

From FAIL to Succeed

The DC FUSION Cohort had to be willing to FAIL where FAIL means First Attempt In Learning. Once Cohort participants accepted the important role of failure and making mistakes in the learning process, learner-centered professional development could begin. The ability to FAIL and discuss failure in reflective videos shared by peers placed all participants on the same level of vulnerability. In one activity, DC FUSION Cohort participants were assigned an activity where they had to go home and video themselves using aluminum foil to conduct electricity in a circuit. The end goal was to build a complete circuit using a light emitting diode (LED) bulb and a 3V lithium battery. The homework instructions were simple. Teachers were directed to state a hypothesis prior to engaging in the act of the experiment. They were instructed to keep the video going, while talking through any mishaps or "eureka" moments. The results of this assignment led to great academic discourse and personal reflection during the following professional development session. Many of the participants felt the video session allowed for greater personal discourse as it related to inquiry and problem-solving. Many participants wanted to know what other everyday materials could be used to conduct electricity. Others thought of ways to utilize this digital approach to engage students in character analysis and to defend math problems. Once FAIL became a platform for growth and possibilities, cohort participants eagerly posted classroom experiences with STREAM activities. Immediate transformation of a STREAM discourse or common STREAM language was established. FAIL posters were developed in classrooms. Many participants from both cohorts of teacher participants added reflective components to enhance the students' overall experience with the lesson content. One participating school leader with guidance from the assigned DC FUSION Coach, empowered their DC FUSION Cohort participants to lead a mini professional development workshop for the faculty at their school on STEM education. This same school would later host a "Family STEAM Night!" In support of this promising outcome, Joyce and Showers (2002) report that teachers need specific attitudes and skills, such as persistence, the ability to use peers productively, understanding of how training is transfered, and an understanding of the need for theory to become more effective learners.

LIMITATIONS, SOLUTIONS, AND RECOMMENDATIONS

Effective professional development programs often require a significant financial investment. The DC FUSION STREAM PD Consortium was able to take money out of the equation for the cohort of urban educators in the District of Columbia as a result of the Teacher Quality Improvement Grant. The value of "Student Achievement" is priceless and it is up to organizations like the DC STREAM FUSION Consortium to continue to make quality professional development available to underserved and underfunded communities. In addition, creating a more robust scaffold for building-level leaders will help ensure professional development sustainability, by incorporating Stage Two of the Professional Development process, "The Teacher Participant as the Peer Coach." One of the most requested components of the TQI PD Grant during post-reflective surveys was to incorporate the assigned coaches into the building more than five days. Both teachers and building level leaders felt that the impact of inserting a coach made the transition from theory to practice an achievable reality.

Daily effort sustained over a long period of time yields Olympic caliber performance. How can this concept be embraced and utilized by our teachers? One very important factor that was mentioned previously involves our aversion to negative assessments even to the point of staying at a level of mediocrity rather than addressing the constructive criticism and improving our skill level. Utilizing professional development as a framework for effective classroom instruction means tackling tough issues such as being unfamiliar with content. Through academic discourse and self-reflection, teachers will be able to adjust practices based on peer interaction, facilitator guidance, and instructional support from classroom coaching.

The end goal of many professional development opportunities is to impact student achievement. The DC STREAM FUSION PD Consortium designed each workshop to help eliminate "Pedagogical Discontentment" through a blended PD approach. By pairing instructional coaching with workshop style professional development, DC FUSION Participants received additional guidance to help in implementing concepts learned. In addition, the assigned coach was able to support DC FUSION Participants in writing lessons, and modeling and introducing additional strategies and concepts to further enhance instructional practice to improve the level of concept mastery for students. Likewise, administrators also needed basic help to support the DC FUSION Participants serving as educators in their buildings.

An additional goal derived from the DC FUSION project was to provide a roadmap through academic discourse and reflection. The notion was to coach administrators on key concepts and to front load them with knowledge of how the DC FUSION Cohort participant's class might transform over the course of the academic year. It was important for the building level administrator to be open to new ideas, concepts, and approaches to presenting content in the classroom.

FUTURE RESEARCH DIRECTIONS

Michael Fullan (2011) identified four factors that are often wholeheartedly embraced to enable large scale education reform. The factors are 1) accountability (rather than capacity building), 2) individual quality (rather than group quality), 3) focus on technology (rather than instruction), and 4) Fragmentation (rather than systematic planning and implementation). A 'wrong driver' is a deliberate policy that has little chance of achieving the desired result. A 'right driver' is one that ends up achieving better measurable results for students. Specifically, the drivers should 1. foster intrinsic motivation of teachers and students; 2. engage educators and students in continuous improvement of instruction and learning; 3. inspire collective or team work; and 4. affect all teachers and students. It is surprising to note that what should be done in education reform is typically known clearly. The challenge comes in how to accomplish what needs to be done by the appropriate design, implementation, and assessment of key interventions in a way that provides unequivocal conclusions.

There are several research questions that can contribute to efforts to pinpoint what needs to be done to develop effective strategies to address challenges in providing the most effective STREAM PD with impactful instructional coaching that addresses all facets of teacher needs for large scale impact. It will be very important to study how analogies in general influence student learning and determine how they can be used most effectively to enhance both teacher and student learning in the rich context of coaching. What is the proper role of coaching in teacher PD and how can it be tailored to have the biggest impact on teacher and student learning? More research is needed to support targeted instructional coaching in the classroom to transform learning experiences. In addition, challenges related to the scale up of effective instructional coaching interventions need to be addressed for the rich benefits of instructional coaching to make its full contribution to transforming STREAM PD for optimized teacher and student learning.

CONCLUDING THOUGHTS

Professional Development is a critical factor in transforming urban classrooms. When designing effective PD's, one must be mindful of the intended outcome and the intended audience and stakeholders. The intended audience includes teachers and administrators within schools with key stakeholders consisting of students who directly benefit from the PD interventions and their parents and communities. Being cognizant of the challenges that exist in designing comprehensive professional development means designing an engaging experience built on a framework to support teachers throughout the year. When teachers are confident in content, student success is inevitable. The design of more effective STREAM PD will have a big impact in enhancing the level of inspiration that will permeate urban classrooms and transform the learning environment. Analogies can be used for deeper learning and can enrich our understanding of instructional coaching and how it can be useful in transforming STREAM PD and teacher and student achievement.

The Teacher Quality Improvement Grant provided a platform for the *DC FUSION STREAM PD Consortium* to begin developing an engaging professional development model for educators in the Washington, DC area. As a result, four universities and STEM and educational professionals have developed a working framework that includes providing instructional coaching as a key element of professional development as it relates to supporting teachers in the classroom. Early results based on the external evaluator report and individual teacher surverys shows that the guided support provided by the instructional coach was a very effective component of the project. Teachers expressed higher levels of confidence in their ability to incorporate technology into their lessons after having the opportunity to create and use it themselves. This was another highlight of the TQI Project. Participants expressed the desire to have more than five one-on-one sessions with the instructional coach. Further study is needed to determine the length of time an instructional coach should spend with participants during the visit and the number of total visits during the academic year.

Many teacher participants expressed a change in their approach to lesson development and lesson presentation in the classroom. These discussions ranged from creating more inquiry-based activities to allow students to seek solutions and to discover and research topics covered under course bsed standards. Other teachers expressed the desire to team with non-participating DC FUSION participants in their school to ensure consistency in teaching was occurring across the grade levels. As the DC FUSION STREAM PD Consortium continues to disaggregate the data, the goal is to adjust and enhance the model based on teacher experiences and direct instruction adjustments based on STREAM strategies that were successfully implemented in the DC FUSION participant classrooms.

ACKNOWLEDGMENT

This research was supported by funds provided by the Teacher Quality Improvement (TQI) Grant Program, authorized under Part A, Subpart 3 of Title II of the Elementary and Secondary Education Act of 1965 (ESEA), as amended, and administered by the District of Columbia Office of the State Superintendent of Education (OSSE - OSSE Grant Number:72367B. Federal grant info: CFDA/SUBPROGRAM NO: 84.367B).

REFERENCES

Aguilar, E. (2013). *The art of coaching: Effective strategies for school transformation*. Hoboken, NJ: John Wiley & Sons.

American Association for the Advancement of Science (AAAS). (1989). Project 2061: Science for All Americans. Washington, DC: AAAS.

American Association for the Advancement of Science (AAAS). (1993). *Benchmarks for Science Literacy*. New York: AAAS.

American Association for the Advancement of Science (AAAS). (2001a). *Atlas of Science Literacy* (Vol. 1-2). Washington, DC: American Association for the Advancement of Science.

American Association for the Advancement of Science (AAAS). (2001b). *Designs for Science Literacy*. New York: AAAS.

Annetta, L. A., Frazier, W. M., Folta, E., Holmes, S., Lamb, R., & Cheng, M. T. (2013). Science teacher efficacy and extrinsic factors toward professional development using video games in a design-based research model: The next generation of STEM learning. *Journal of Science Education and Technology*, *22*(1), 47–61. doi:10.100710956-012-9375-y

Barkana, R. (2018). Possible interaction between baryons and dark-matter particles revealed by the first stars. *Nature*, *555*(7694), 71–74. doi:10.1038/nature25791 PMID:29493590

Bennett, A., Bridglall, B., Cauce, A., Everson, H., Gordon, E., Lee, C., ... Stewart, J. (2004). *All Students Reaching the Top: Strategies for Closing Academic Achievement Gaps. A Report of the National Study Group for the Affirmative Development of Academic Ability*. Naperville, IL: Learning Point Associates.

Boaler, J. (2016). *Mathematical mindsets: Unleashing students' potential through creative math, inspiring messages and innovative teaching*. John Wiley & Sons.

Bowman, J. D., Rogers, A. E. E., Monsalve, R. A., Mozdzen, T. J., & Mahesh, N. (2018). An absorption profile centred at 78 megahertz in the sky-averaged spectrum. *Nature*, *555*(7694), 67–70. doi:10.1038/nature25792 PMID:29493587

Darling-Hammond, L., & Bransford, J. (Eds.). (2007). *Preparing teachers for a changing world: What teachers should learn and be able to do*. John Wiley & Sons.

Darling-Hammond, L., & McLaughlin, M. W. (1995). Policies that support professional development in an era of reform. *Phi Delta Kappan, 76*(8), 597–604.

Editorial, N. P. (2014). The Power of Analogies. *Nature Photonics, 8*(1), 1. doi:10.1038/nphoton.2013.359

Ertmer, P. A., Richardson, J., Cramer, J., Hanson, L., Huang, W., Lee, Y., ... Um, E. J. (2005). Professional development coaches: Perceptions of critical characteristics. *Journal of School Leadership, 15*(1), 52–75. doi:10.1177/105268460501500103

Fullan, M. (2011). *Choosing the Wrong Drivers for Whole System Reform. Seminar Series Paper No. 204*. Victoria, Australia: Centre for Strategic Education.

Gordon, E. W. (2001). *Affirmative Development of Academic Abilities. Pedagogical Inquiry and Praxis. Number 2*. New York, NY: Columbia University.

Grant, A. (2013). The Efficacy of Coaching. In J. Passmore, D. B. Peterson, & T. Freire (Eds.), *The Wiley-Blackwell Handbook of the Psychology of Coaching and Mentoring* (pp. 15–39). Hoboken, NJ: John Wiley & Sons.

Green, L. S. (2014). Through the looking glass: Examining technology integration in school librarianship. *Knowledge Quest, 43*(1), 36–43.

Joyce, B. R., & Showers, B. (2002). *Designing Training and Peer Coaching: Our needs for learning*. ASCD.

Koppal, M., & Caldwell, A. (2004). Meeting the Challenge of Science Literacy: Project 2061 Efforts to Improve Science Education. *Cell Biology Education, 3*(1), 28–30. doi:10.1187/cbe.03-10-0016 PMID:22039343

Kowal, J., & Steiner, L. (2007). *Instructional Coaching. Issue Brief.* Center for Comprehensive School Reform and Improvement.

Kraft, M. A., Blazar, D., & Hogan, D. (2018). The effect of teacher coaching on instruction and achievement: A meta-analysis of the causal evidence. *Review of Educational Research, 88*(4), 547–588. doi:10.3102/0034654318759268

Marzano, R. J., Simms, J. A., Roy, T., Heflebower, T., & Warrick, P. B. (2013). *Coaching classroom instruction*. Bloomington, IN: Marzano Research Laboratory.

Matlen, B. J., Vosniadou, S., Jee, B., & Ptouchkina, M. (2011). Enhancing the comprehension of science text through visual analogies. In L. Carlson, C. Holscher, & T. Shipley (Eds.), *Proceedings of the 34th Annual Conference of the Cognitive Science Society* (pp. 2910–2915). Academic Press.

MET Project Research Paper. (2010). *Learning About Teaching: Initial Findings from the Measures of Effective Teaching Project.* Bill and Melinda Gates Foundation.

Nadelson, L. S., Seifert, A., Moll, A. J., & Coats, B. (2012). i-STEM summer institute: An integrated approach to teacher professional development in STEM. *Journal of STEM Education: Innovation and Outreach.*

National Education Association. (2012). *Preparing 21st century students for a global society: An educator's guide to the "Four Cs"*. Alexandria, VA: National Education Association.

National Research Council. (2000). *How People Learn: Brain, Mind, Experience, and School.* Washington, DC: National Academy Press.

National Research Council. (2011). *Successful K-12 STEM education: Identifying effective approaches in science, technology, engineering, and mathematics*. National Academies Press.

National Research Council. (2012). *A Framework for K-12 Science Education: Practices, Crosscutting Concepts and Core Ideas.* Washington, DC: National Academy Press.

National Research Council. (2013). *Next Generation Science Standards: By States for States*. Washington, DC: National Academy Press.

National Science Foundation. (2014). *Science & Engineering Indicators.* National Science Board. NSB 14-01.

National Staff Development Council (US). (2001). *National Staff Development Council's Standards for Staff Development.* National Staff Development Council.

Pollack, J. (2015). *Shortcut: How Analogies Reveal Connections.* Spark Innovation, and Sell Our Greatest Ideas.

President's Council of Advisors on Science and Technology (US). (2010). *Prepare and Inspire: K-12 Education in Science, Technology, Engineering, and Math (STEM) for America's Future: Executive Report.* Executive Office of the President, President's Council of Advisors on Science and Technology.

Puentedura, R. R. (2014). *Frameworks for Educational Technology: SAMR, the EdTech Quintet, and the Horizon Report.* Paper presented at the International Conference on Teaching and Learning with Technology (iCTLT) Conference, Singapore.

Vendetti, M. S., Matlen, B. J., Richland, L. E., & Bunge, S. A. (2015). Analogical reasoning in the classroom: Insights from cognitive science. *Mind, Brain and Education: the Official Journal of the International Mind, Brain, and Education Society*, *9*(2), 100–106. doi:10.1111/mbe.12080

Wang, M. C., & Kovach, J. A. (1995). *Bridging the Achievement Gap in Urban Schools: Reducing Educational Segregation and Advancing Resilience--Promoting Strategies*. Publication Series# 95-9.

Will, M. (2018). *'An Expensive Experiment': Gates Teacher-Effectiveness Program Shows No Gains for Students*. Retrieved from https://www.edweek.org/ew/articles/2018/06/21/an-expensive-experiment-gates-teacher-effectiveness-program-show.html

Williams, B. (Ed.). (1996). *Closing the achievement gap: A vision for changing beliefs and practices*. ASCD.

Zuger, S. (2012). The state of STEM (science, technology, engineering, and mathematics). *Technology and Learning*, *33*(3), 38–44.

Chapter 5
Bolstering Content by Promoting Language and Literacy in the Urban Science Classroom

Jennifer E. Renn
Purdue University, USA

Annie Laurie Duguay
Center for Applied Linguistics, USA

ABSTRACT

This chapter focuses on the creation and implementation of lesson modules related to developing language skills in the middle school science classroom. These modules, which focus on academic language development for students who are English learners or speakers of non-standard dialects, are part of the curriculum in a Master of Arts in Teaching (MAT) program for middle school science educators at a large urban U.S. university. Drawing on the literature surrounding the academic language of science and language development, the content of these modules is appropriate for teacher educators and science teachers at all levels. The module content was designed around three components, linguistic understandings, critical language awareness, and pedagogical modeling, and exemplifies the Standards of Professional Development Excellence established by the Center for Applied Linguistics. This chapter describes the module creation, observations from the module delivery, and program participant responses to the materials and content.

DOI: 10.4018/978-1-5225-7814-7.ch005

INTRODUCTION

Despite the growing diversity in urban U.S. classrooms, most pre-service and in-service content teachers receive little to no training on how to work effectively with linguistically and culturally diverse students who are in need of linguistic support. This lack of preparation often results in teachers approaching students who are in the process of acquiring English as an additional language (also referred to as emergent bilingual students, English learners, or ELs) from a deficit perspective rather than leveraging the knowledge, skills, and abilities that students bring to the classroom (Valencia, 1997). Attitudes are often similar for speakers of non-standard varieties of English, including African American English, where teachers make assumptions about students' abilities based on their language (Rickford, 2005; Alim, 2005). Rickford (2005) notes the parallels between students who speak a language other than English at home and students who speak a non-standard variety of English at home. In both cases, he recommends viewing the language these linguistically and culturally diverse students bring to the classroom as a resource in engaging with content concepts.

Viewing linguistically and culturally diverse learners from a deficit perspective can have serious negative long-term consequences for these students. Research suggests that one way in which this has manifested over the last few decades is through the representation of students of color in special education programs (Counts, Katsiyannis, & Whitford, 2018). While studies have shown that minority students are often overrepresented in certain special education categories (Artiles, Harry, Reschly, & Chinn, 2002; Waitoller, Artiles, & Cheney, 2010; Sullivan, 2011), in some cases they may also be under identified for receiving needed services (National Education Association and National Association of School Psychologists [NEA/NASP], 2007). For example, Shifrer, Muller, and Callahan (2011) found that EL students who had caregivers that spoke a language other than English were less likely than those with English-speaking parents to be identified for special education services. Since teacher referral is a primary factor in determining student placement in special services, it is critical that teachers are making recommendations based on an educated assessment of students' abilities and needs, distinguishing linguistic variation from disability or behavioral issues.

In addition, research has shown that preparing teachers to work within a multicultural classroom has a positive impact on both students and teachers (Sleeter & Owuor, 2011). Reviews by Hollins and Guzman (2005) and Sleeter (2008) found evidence that courses designed to prepare teachers to work with diverse students resulted in positive changes of attitudes and gains in cultural consciousness, especially when the courses contained activities that incorporated self-reflection. Such changes in teacher attitudes and perceptions trickles down to impact students. Studies show that students perform better academically when their teachers actively seek to build relationships with them and their families (e.g., Bishop, Berryman, Cavanagh, & Teddy, 2009; Hawkins, Guo, Hill, Battin-Pearson, & Abbott, 2001). In fact, Gay's (2010) review of the literature found that even curricula with minimal culturally relevant content was linked to higher test scores, higher grade point averages, and increased levels of student engagement and self-confidence.

This chapter describes the development, implementation, and effectiveness of content modules on the learning and teaching of linguistically and culturally diverse students that were designed with these concerns in mind. The modules were created for pre- and in-service teaching fellows enrolled in a Master of Arts in Teaching (MAT) program at a public urban U.S. university[1]. Program fellows had a background in STEM, mostly in science, and were committed to teaching middle grades science in an urban school district as part of their program obligations. To meet the needs of the fellows and their students, the modules sought to increase the program fellows' "pedagogical language knowledge," which advocates a "shift in the definition of teaching practice for teachers and teacher educators away from 'English learners' toward 'language use for academic purposes'" (Galguera, 2011, p. 85). This view is further expanded by Bunch (2013), who investigates how to best prepare mainstream content teachers to work with linguistically and culturally diverse students. Bunch contends rather than characterizing them as "language teachers," content teachers must "purposefully enact opportunities for the development of language and literacy in and through teaching the core curricular content, understandings, and activities that teachers are responsible for (and, hopefully, excited about) teaching in the first place" (p. 298). It is only by promoting this perspective that teachers are able to effectively prepare *all* students to meet the increased language and literacy demands that have been incorporated into standards-based instruction. Given the linguistic diversity of the urban U.S. classroom, this perspective is especially relevant to K-12 urban STEM educators and is therefore an integral component of the program modules.

The chapter that follows begins with a review of relevant literature on challenges faced by linguistically and culturally diverse students in the STEM classroom to highlight the need for linguistic content in K-12 educator training programs, followed by an overview of the characteristics of effective professional development (PD) for content teachers. Next, it describes the development of content modules on academic language development that were designed for pre- and in-service middle grades science teachers, who were enrolled in the urban MAT program mentioned above. The explanation of content modules is organized according to the Standards of PD Excellence established by the Center for Applied Linguistics (forthcoming) in order to illustrate their alignment with these best practices. The chapter then discusses feedback on the modules from program participants and instructors and concludes with recommendations and next steps.

BACKGROUND

Linguistically and Culturally Diverse Students and STEM

Classroom culture in the United States tends to align with the beliefs and practices of European American households (Rogoff, 2003), even in urban areas that serve large numbers of minority students, including African Americans and ELs. Consequently, linguistically diverse students' language practices have often been misunderstood in school settings. The "home-school mismatch" theory explains one of many reasons that linguistically diverse students struggle in school (Heath, 1982, 1983; Michaels, 2006; Philips, 1983). Students are often assessed, formally and informally, by the ways they use language in school settings; if students are not able to articulate understandings in culturally recognizable ways in classrooms, they may be assessed inaccurately. For EL students, their level of English language proficiency may make it more difficult for them to engage with content material without support, but language issues may still arise for native English speakers as well. For example, some researchers have hypothesized that, for some African American students, dialect may play a role in this disparity because of a mismatch between the varieties of speech that many African American children use at home and the language used in school. A student who is not proficient in Standard American English (SAE) may experience difficulty with a curriculum based on SAE, particularly with regard to literacy (Rickford, 1999). This is especially relevant for speakers of African American English (AAE) because of the widespread use and diverse nature of this dialect. This situation has an enormous impact on the quality and effectiveness of the education received by speakers of nonstandard dialects like AAE.

These linguistic and cultural differences may be a factor in the achievement outcomes of minority students in both classroom-level and large-scale assessment data. Given the reading and writing demands in both science content work and science assessments, students' language and literacy skills play a big role in student outcomes. Student scores on the National Assessment of Educational Progress (NAEP), for example, highlight the impact of current educational practices on the academic performance of diverse learners. Figure 1 shows the results of the NAEP reading assessment (2015; 2017), which has 500 possible points. On this assessment, the average African American score was 26 points lower than their European American peers in Grade 4 and Grade 8, and 29 points lower in Grade 12. The discrepancy was even greater for ELs, who scored 37 points lower than their non-EL peers in Grade 4, 43 point lower in Grade 8, and 49 points lower in Grade 12.

Figure 1. Scale scores on the NAEP reading assessment in Grades 4, 8, and 12 by group

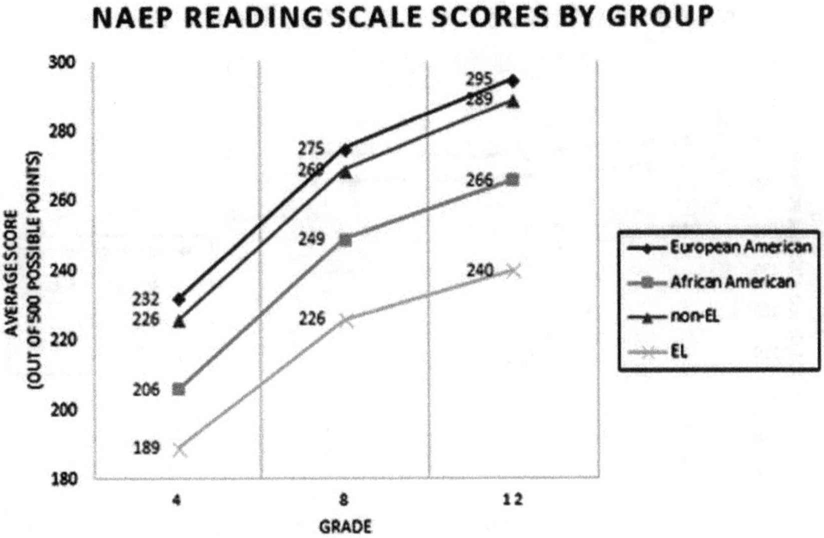

Figure 2 shows that on the NAEP writing assessment (2002; 2011), which has a total of 300 possible points, these gaps showed a larger increase over time, with African American students, on average, scoring 21 points lower than their European American classmates in Grade 4, 26 points lower in Grade 8, and 29 points lower in Grade 12. Once again, the average EL student experienced even greater disparities, scoring 28 points lower in Grade 4, 44 points lower in Grade 8, and 56 points lower than their non-EL peers in Grade 12. Also noteworthy is the steep decline in writing scores for EL students with age, whereas the other groups' scale scores are more stable across grades.

These trends persist in content area assessments. The results of the NAEP science assessment (2015), shown in Figure 3, are similar to results for reading and writing, but differences between groups are even larger. In the science assessment, which also has a possible 300 points, the average African American score was 33 points lower than their European American peers in Grade 4, 34 points lower in Grade 8, and 35 points lower in Grade 12. The discrepancy was even greater for ELs, who scored 37 points lower than their non-EL peers in Grade 4; this increased to a 47-point discrepancy in Grade 8 and Grade 12. These results demonstrate that not only do European American and non-EL students outperform their African American and EL peers in measures of reading and writing, but also in many cases, these differences grow over time, highlighting the critical need for educators to connect and engage with students from diverse backgrounds.

Figure 2. Scale scores on the NAEP writing assessment in Grades 4, 8, and 12 by group

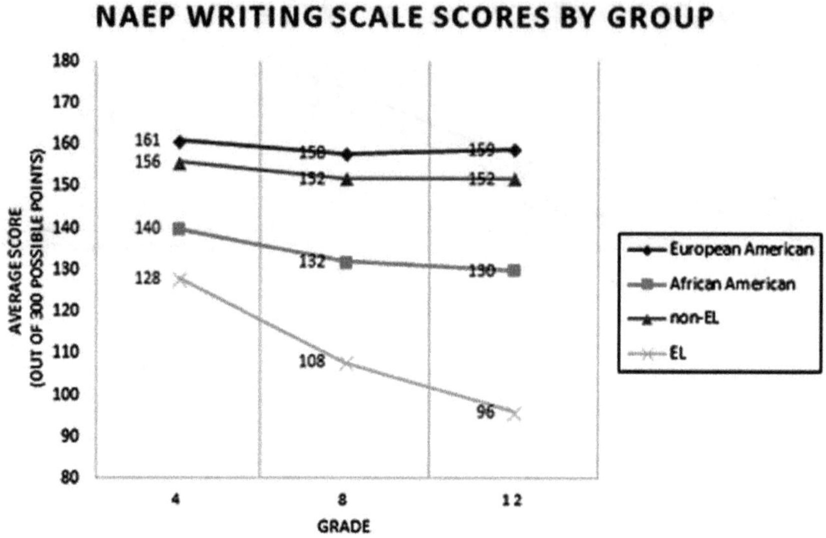

Figure 3. Scale scores on the NAEP science assessment in Grades 4, 8, and 12 by group

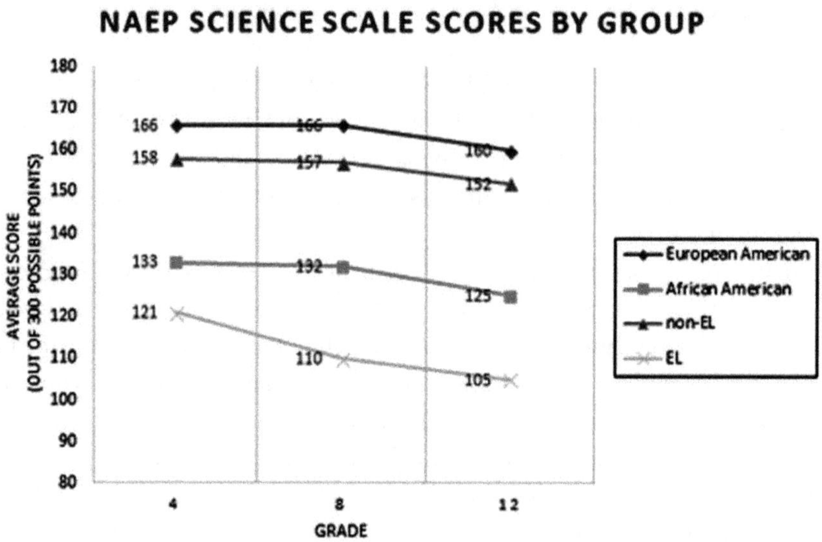

In addition, when teachers do not recognize and support the skills of linguistically and culturally diverse learners, students often become disengaged from classroom content making them at risk for drop-out. Research on linguistically diverse students' scientific discourse has shown that diverse students can engage in productive scientific discourse when they are taught in ways that value and take their resources into account (Hudicourt-Barnes, 2003). Warren and colleagues (2001), for example, have focused on capitalizing on the congruence between students' and professional scientists' habits and practices. By identifying key "points of contact" between students' ways of knowing and scientific ways of knowing, they have successfully promoted students' adoption of scientific discourse. Thus, a key linguistic understanding is that all students' ways of using language in the classroom should be valued, understood, and leveraged. Past experimental research on science instruction for ELs demonstrates that curriculum and instructional techniques that have proven to be successful for monolingual students, such as inquiry-based learning, are effective but not sufficient for the achievement of ELs. Instruction that incorporates ELs' language and literacy needs as well as addresses their cultural backgrounds is more effective (August, Artzi, & Mazrum, 2010; Lee, Maerten-Rivera, Penfield, LeRoy, & Secada, 2008). Examples of modifications for ELs when integrating content and language instruction include visuals, explicit vocabulary and morphology instruction, interactive reading, and first language glossaries (Echevarría, Vogt & Short, 2017).

In an era of rigorous standards and assessment, there are multiple language and literacy demands in STEM content standards. The Framework for English Language Proficiency Development Standards Corresponding to the Common Core State Standards and the Next Generation Science Standards (Council of Chief State School Officers, 2012) outlines the *practices*, or disciplinary tasks and activities that are characteristics of each content area. Each of these disciplinary-specific practices includes a related language and literacy practice. For example, constructing viable arguments in math may include listing the steps of a geometry proof or justifying the answer to a word problem; in science, engaging in arguments from evidence may include recording daily observations during a lab or orally describing the data trends with implications for that subject area. Explicit instruction on these language practices, including the language specific to each content area and the connections to students' home languages and home cultures, is vital to the achievement of linguistically and culturally diverse students in science (Quinn, Lee, & Valdés, 2012).

Specifically, the content area of science has its own academic language at the discourse, sentence, and word or phrase levels of language. In fact, each discipline within science (e.g., chemistry, physical science, biology, etc.) has its unique linguistic features (Quinn, Lee, & Valdés, 2012). Discourse, the ways in which we are socialized to communicate through language, is the broadest level of language. For example, in science, students need to be able to attend to precision in measurement, note cause and effect language, and assess the validity of scientific claims. At the sentence level, the grammatical level of language, certain structures of language are highly frequent in science texts, such as nominalization of verbs and adjectives to create abstract noun phrases (e.g., the *rotation* of the planets). Additionally, passive voice is used to maintain objectivity and timelessness. In terms of the word or phrase level, typically the most instructed level of academic language, technical vocabulary with highly specialized meanings is used to "pack" information, creating a high lexical density. Words that are used in everyday language, such as *table*, *force*, and *cell*, are used with unique specificity in science. Furthermore, texts might use phrases such as "the authors claim" or "the results were confirmed" which contain a subtle evaluation of the findings, requiring a nuanced understanding of language than typically used in everyday conversation (Halliday & Martin, 1993; Schleppegrell, 2004). Lastly, students need to become experts in non-linguistic interpretations of diagrams, equations, charts, and laboratory equipment. Students need to be taught these features of language in order to unpack meaning in science.

Effective STEM Teacher Preparation

The aim of the MAT module content was to develop the linguistic understandings of teachers to better integrate language into their instruction of science concepts. Preparing educators to work effectively with diverse learners in the STEM classroom is a key component to ensuring student success in urban classrooms. Content-focused professional development (PD) was found to improve teachers' knowledge of the subject area as well as content-specific pedagogy and aspects of instructional practices (National Center for Education, Evaluation and Regional Assistance, 2016). The National Research Council (NRC), summarizing decades' worth of research on effective STEM education practices, state that "effective instruction capitalizes on students' early interest and experiences, identifies and builds on what they know, and provides them with experiences to engage them in the practices of science and sustain their interest" (NRC, 2011: 18). For linguistically and culturally diverse students, building on what students know includes their first language and culture. The Committee on a Conceptual Framework for New K-12 Science Education Standards concluded that "Ultimately, the interactions between teachers and students in individual classrooms are the determining factor in whether students learn science successfully. Thus, teachers are the linchpin in any effort to change K-12 science education…[and]…the initial preparation and professional development of teachers of science will need to change" (NRC, 2012: 255). Moreover, students who worked with more effective science teachers exhibited higher performance on the state high school science exam (Johnson, Zhang, & Butler Kahle, 2012). A review of the research, specifically on ELs' achievement in science, suggests "that effective first language instruction, most notably inquiry-based learning, may be a good starting place, but it is not sufficient to help English language learners learn science.... Interventions that build on effective first-language science teaching research, but also take into account the language and cultural backgrounds of English language learners may be more promising" (August et al., 2010: 1).

One way to address the needs of linguistically and culturally diverse students in the classroom is through sheltered instruction, such as the SIOP Model (Echevarría, Vogt, & Short, 2017). With eight components and 30 features related to integrating content and language instruction, the SIOP Model is a lesson preparation and delivery framework that helps educators to consider the language demands of a content topic or text and scaffold those domains through techniques such as pre-teaching vocabulary, providing first language support, and promoting student interaction. In fact, in science classrooms with teachers who demonstrated a high level of implementation of the SIOP Model features, students made more significant gains on a science post-assessment (Echevarría, Richards-Tutor, Chinn, & Ratleff, 2011). Despite the evidence behind such effective practices, research has shown that the needs of linguistically diverse students and their teachers have been overlooked, and that teachers feel unprepared to meet their English learner students' needs, particularly in the implementation of rigorous content area standards (Gándara, Maxwell-Jolly, & Driscoll, 2005; Quinn, Lee, & Valdés, 2012). Research has shown that teachers who have had more in-service and pre-service PD on the topic of language learners have greater confidence in their ability to teach these students effectively (Gándara, Maxwell-Jolly, & Driscoll, 2005). With the increasing diversity of US school populations, rigorous college and career standards, and the fact that language and literacy are embedded in all grade-level content topics and tasks, teachers need to feel confident and knowledgeable about language, including its use and instruction (Lindahl & Watkins, 2015; Wong Fillmore & Snow, 2018).

URBAN EDUCATOR PREPARATION PROJECT

Given the need for increased educator training to support linguistically and culturally diverse students, especially in urban settings, this paper describes the development and implementation of language and literacy-related content that was integrated into a MAT program at a public urban university. This NSF-funded program addresses effective teacher preparation to reach linguistically and culturally diverse students by integrating science content knowledge, the pedagogical practices of inclusive instruction, and the attention to language and literacy as embedded in science communication. Experts from a non-profit that focuses on access and equity for linguistically and culturally diverse students created modules for face-to-face and online class sessions that were delivered by university faculty for educators who will teach science to middle grades students in a large urban school district. The experts developed the modules based on their work in urban schools throughout the country and initially designed the content for educators in both math and science; since the MAT program fellows were science educators, however, the materials were adapted to focus on language and pedagogy in the science classroom. They then piloted the material, observed the implementation of the modules in the MAT program, and modified and recorded the content for sustained implementation.

The project module content was developed for the fellows in the MAT program, but the content was first pilot-tested with science educators from the same urban school district where the program fellows would complete their teaching assignments. The science educators, some of whom later acted as mentors to the teaching fellows, were asked to evaluate the content for its appropriateness in their diverse grade-level classrooms. In order to build capacity for the university faculty, research team recorded the modules and held workshops to train the MAT professors, who would be teaching the content to the fellows, to deliver the scope and sequence of the program. Additionally, the experts attended all of the in-class presentations of the modules to gather feedback on implementation. The linguistics experts worked particularly closely with the lead faculty for the science pedagogy course to revise modules based on her instruction, student assessment, and the ease of delivering the materials.

The MAT module content was designed around three components: linguistic understandings, critical language awareness, and pedagogical modeling. Three foundational *linguistic understandings* are emphasized in the program content: (1) that all students' existing styles of language use and home language are valued and treated as resources in classrooms; (2) that students need to develop an awareness of Academic English of science; and (3) that the Academic English of science should be simultaneously developed alongside conceptual understanding of science ideas. Each session was also tailored to promote *critical language awareness* (CLA). Fairclough (1992) defines critical language awareness as, "how language conventions and language practices are invested with power relations and ideological process which people are often unaware of" (p. 7). Throughout the module content, language is not viewed from a form-based perspective but instead is treated as a way to make meaning in a social context; those contexts are always associated with ideologies. Alfouaim (2012) writes, "[language] educators who are armed with CLA knowledge are capable to change the conventional ideology that proposes the superiority of one language over another for any reason." Because language is so central to classroom learning, it is important for teacher candidates to critically reflect on the role it plays in teaching and learning. Finally, the module content was designed to *model pedagogical approaches* to teaching linguistically diverse students. Strategies and best practices are embedded into module activities so that teacher candidates experience first-hand the types of activities that may be appropriate for teaching linguistically and culturally diverse students. Each session is intended to model a different type of pedagogical activity aimed at promoting academic language development with linguistically diverse students.

The project content, including online modules and professional development sessions, also exemplifies the Standards of PD Excellence put forward by the Center for Applied Linguistics (forthcoming). These standards refer to the teacher professional development of educators of linguistically and culturally diverse students for both in-service and pre-service educators. Based on these standards, the PD sessions for teacher candidates in the MAT program and for their professors were:

1. Research-based, content driven, and relevant;
2. Meaningful and intellectually stimulating;
3. Engaging, interactive, and collaborative;
4. Well-organized and facilitated;
5. Positively-framed, respectful, and inclusive; and
6. Supportive of future learning and growth.

1. Effective Professional Development Is Research-Based, Content-Driven and Relevant

Teacher preparation for educators of linguistically and culturally diverse students should be grounded in current research on language and literacy acquisition, pedagogy, content learning, and culturally-relevant instruction, particularly in a master's program. In addition, PD should be content-driven and relevant to what the educators need to know to meet the needs of their linguistically and culturally diverse students, including information about language (Wong Fillmore & Snow, 2018). According to the Center for Applied Linguistics (forthcoming), PD should be relevant to the population and needs of the students, including their grade levels, language and content proficiency, immigration stories, educational backgrounds, as well as dialects and native languages. The PD should also be responsive to the program models, community needs, and curricula of instruction.

This standard was met by the project in a variety of ways. The experts included two professionals with doctorates in linguistics and education and a director of professional development with middle grades instructional experience in a science-focused urban school. All of the language experts had experience with classroom interventions, specifically in the science content area. Additionally, the assigned articles and homework tasks were research articles from the areas of linguistics, language learning, and instruction for linguistically and culturally diverse students.

In terms of student relevance, due to the fact that the non-profit is based in the same large urban area, the experts' familiarity with the local schools, community makeup, and the university itself allowed them to tailor the materials to the needs of students and educators in the city. More significantly, the experts are knowledgeable about the linguistic variation in the community, including the predominance of African American English, which was of particular importance given the demographics of the local school district. In addition, the module materials were piloted first to the mentor science teachers from the urban district to ensure student relevance. Upon their response that the types of activities and approaches to integrating content and language would be useful in their classrooms, the materials were deemed to be relevant to this school population.

To ensure that the project materials and trainings were content-driven, the experts drew from the scope and sequence of the MAT program and the focal science areas each semester (physical vs. biological sciences). Since the focus of the content of this portion of the MAT program was language and language learning, the experts included lessons on the three levels of language (i.e., discourse, sentence, and word/phrase) and discussed classroom implications. For example, participants did a chart puzzle activity with the three levels of language (see Figure 4) to match the linguistic features of academic language (e.g., language functions or grammatical metaphors) with definitions and examples of each feature. The activity was intended to create linguistic awareness rather than to have the participants memorize the definitions of linguistic terms.

Following this awareness-building, the content of the courses focused on ways to apply this knowledge of the three levels of language in the classroom. While one could not imagine middle school students doing a puzzle of these linguistic definitions in a science classroom, teaching fellows were asked to generate ideas about: 1) *why* the puzzle activity would benefit linguistically and culturally diverse students (i.e., gives students a low-anxiety way to interact with peers on challenging grade-level material; allows participants to speak, listen, and read; contains a kinesthetic element; provides a more engaging way to read a chart, etc.), and 2) *how* they could use this same activity in their classrooms (e.g., by cutting up a chart with information on biomes or the periodic table and having students reassemble the charts in groups as a puzzle). This emphasis on classroom implications throughout the coursework was appreciated by the teaching fellows as a model for their own classrooms with time to plan their own engaging activities. While K-12 students or even science teachers would be expected to define each feature of academic language, the modules provided applications for such knowledge. For example, while many teachers focus on vocabulary, the language impeding comprehension for linguistically diverse students may be found at a grammatical or discourse level of language. Therefore, if teachers are attuned to analyzing the academic language of a text, video, or lesson activity, they can select a small language feature for more focused instruction that may help students better understand and communicate a science concept. For example, in Figure 5, the experts focused not only on the listening vocabulary related to gravity, but also to the key verbs used to describe how scientists make predictions and subsequently validate their findings (e.g., *revealed, convinced, contradicted, reasoned*, etc.).

In addition to the content and student relevance, the experts also focused on the relevance of the materials format. Based on the hybrid nature of the courses (i.e., partially online, partially face-to-face), the experts drafted PowerPoint slides with notes. The notes were recorded through PowerPoint and clear instructions were given for supplementary handouts with naming conventions. Adjustments were made to the directions or materials based on how well they were received by the faculty member or the students.

2. Effective Professional Development Is Meaningful and Intellectually Stimulating

Like any learning opportunity and in accordance with adult learning theory, teacher preparation for educators of linguistically and culturally diverse students should offer continuous opportunities to use higher order thinking skills in all aspects of the PD sessions (Zepeda, 2012). PD should encourage participants to reflect on their practices and make positive changes in their planning, instruction, and assessment techniques. The PD should model effective pedagogical strategies that can be used immediately and directly with students for teachers that have very limited time to plan lessons. Effective PD should also offer opportunities for practice and application of relevant and authentic content utilizing grade-level appropriate examples.

This standard was met through a variety of techniques, including modeling of teaching strategies, offering opportunities for peer-led discussions, hands-on activities, scaffolded planning time, video lessons of grade-level science instruction, and wait time for processing information. After each modeled strategy, participants had an opportunity to name the reasons why that activity would meet the needs of

Figure 4. Handout that accompanied a matching puzzle of the linguistic feature, definition, and examples

Handout 2.B
Module 1
Session 2: Linguistic Aspects of Academic Language

Academic Language at Three Levels of Linguistic Generality

Discourse Level

- The discourse level of language is language beyond the sentence in both written and oral form. It includes:

Language functions	The communicative purpose of language (e.g., explaining, comparing, sequencing)
Genres	The ways of using language (e.g., narratives, arguments, procedural)
Speech events	Activities with rules or norms for the use of speech (e.g., oral reports or debates)
Reference	The relationship between expressions of spoken language or text or to objects in the real world (e.g., *"that"* or "I don't like *them*").
Cohesion	Linguistic features which tie ideas together (e.g., "conversely", "meanwhile", "because").

Sentence Level

- Sentence level features include different types of grammatical structures and syntax associated with sentence level meaning. It includes:

Grammar	Using a wide variety of grammatical resources (e.g., past tense: he *walked/ran* to the store; perfect: he *has gone* to the store).
Sentence Type	Using simple, compound and complex (e.g., *Rob cooks; Rob cooks and Sara does the dishes; Sara does the dishes only when Rob cooks*).
Grammatical Metaphor	Substituting one grammatical class for another (e.g., We will *observe* the plants over the next two weeks. *Observations* will be recorded in your journals.)

Word/Phrase Level

- Vocabulary includes words and phrases used in social and academic settings. It includes:

General everyday vocabulary	Words that are used in everyday settings (e.g., *book, table, hi*); Note: these words depend on students' varied everyday experiences
General academic vocabulary	Words that are commonly used in academic settings across content areas (e.g., identify, evidence, factor).
Content specific vocabulary	Terms used in particular different content areas (e.g., *Pythagorean, phylum, nationalism*).
Multi-word expressions	Commonly used expressions and sequences of words, including phrasal verbs and transition terms (e.g., *face time, fast food, rely on, based on...*)

linguistically and culturally diverse students in the classroom. This debrief allows teaching fellows to internalize the theory behind the activities and the necessity of that level of scaffolding for English learners and to not get distracted by how engaging they were. While some examples were direct texts and video examples drawn from middle school science curricula (e.g., a reading on invasive plants; a video on gravity), one way the project made materials intellectually stimulating was to incorporate challenging readings on linguistic theory. While some of the content

Figure 5. Listening observation guide, which demonstrates and effective way for linguistically diverse students to watch a video and listen with comprehension. This is an example of an activity the teaching fellows completed that centered on grade-level science content, a video on gravity.

Gravity

Part I: Watch the video. Check off the vocabulary words as you hear them.

Milky Way Galaxy	√	Cosmological constant		15 billion years	
universe		Theoretical physics		Force of gravity	
observations		contradict		Rate of expansion	
static		discoveries		accelerating	
Albert Einstein		Big bang universe		impact	
Force		Dense		Gravity's pull	
rappel		Theory of relativity		astronomers	
repulsive		expansion			

Part II: Find a partner. Decide who will record answers for odd numbers and who will record for even numbers. Circle the answers you are responsible for and read those statements before watching. Watch the video a second time taking notes only for the statements you circled.

1. A hundred years ago, astronomers believed that our Milky Way galaxy was...
2. Observations revealed that the shape and size of the universe was...
3. Albert Einstein believed that...
4. Einstein reasoned that there was a repulsive force, counteracting gravity...
5. Einstein called his "mistake" ____________________.
6. A new wave of discoveries contradicted Einstein's theory began in ______ and by 1929, most astronomers were convinced that the universe...

was challenging for adults who specialize in the sciences, such as the analysis of a text for the three levels of academic language (see Figure 6) the materials were scaffolded to model the type of support that linguistically diverse students would need for rigorous grade-level content. By having the adults negotiate meaning over complex text, comparisons could be drawn to the language and literacy scaffolds needed for their students. In addition, fellows were introduced to some of the more complex readings by completing an extended anticipation guide, shown in Figure

7. This scaffolding technique models the link between oral language and reading comprehension. Before reading an assigned article, teaching fellows read each statement and signaled their agreement. This was followed by discussing their answers with a peer. This prediction process previews the content and language of the article and focuses their attention as they read. Finally, the program fellows signaled their agreement as a post-reading activity and cited their justification for changing their answers or not based on the authors claims. This pre- and post- reading activity is a simple way for K-12 students to bridge from their prior knowledge and their own language to the ideas and language of the text.

3. Effective Professional Development Is Engaging, Interactive, and Collaborative

In effective PD sessions, participants are actively engaged in discussions by questioning, critiquing, reflecting, and planning instruction. PD participants should have opportunities to interact with a variety of participants through cooperative learning structures. To meet this standard, PD should offer participants opportunities to collaborate with one another on a variety of tasks, such as lesson planning and creating classroom activities. PD should also include a variety of activities that utilize varied grouping configurations and all four language domains (reading, writing, listening, and speaking).

Student engagement is particularly important for linguistically and culturally diverse students, as language learning is a social endeavor (Lightbown & Spada, 2006), and the more engaging or intellectually stimulating a task is, the more language is needed to communicate about that topic. Interactive activities were embedded in each MAT module as a model of the types of classroom activities that the teaching fellows should prepare in their classrooms. One way the modules fostered engagement was to incorporate activities that required teaching fellows to use all four language domains. For example, in collaboration with their peers, the fellows participated in guided listening stations, jigsawed readings, answered questions in round robin writing activities (see Figures 8 and 9), and rotated partners in Inside/Outside circle discussions. In creating the modules, the teaching fellows were assigned articles meant to inspire discussions and reflection. The content of these articles was explored through activities like a traveling jigsaw (see Figures 10 and 11), where groups of fellows were responsible for summarizing an assigned article in a poster. Fellows then rotated around the room to read all of the posters and took turns presenting the content to their peers. In addition, to model the fact that more ambiguous or contentious ideas inspire more language, quotes were also chosen to inspire critique and foster opposing viewpoints (see Figures 12 and 13). Activities such as Talking Chips (Kagan, Kagan, & Kagan, 2016) were used to

Figure 6. Teaching fellows analyzed a complex science text for the three levels of academic language and discussed how which linguistic feature may have been most challenging for students and therefore worthwhile for instruction.

Answer Key 2.B
Module 1
Session 2: Linguistic Aspects of Academic Language

Answer Key: Analysis of the Three Features of Academic Language in the *California Invasive Plant Inventory*

- **What kind of discourse does the text represent (is there an identifiable genre or text type)?**
    - This is a website disseminating information; it details the process of identifying and classifying invasive species; it lists results
- **What references are made between spoken language or text or to objects in the real world?**
    - The authors use "its" to refer to plants and "them" to refer to assessments. It also references the "criteria system described below" and the "definitions below"—two references to the same criteria, but using different terms.
- **How are ideas connected? Are there words or phrases that are used to connect ideas?**
    - The text uses transitional terms like *although, therefore*, and *conversely*
- **What language functions are expressed?**
    - explaining
    - categorizing
    - listening
- **What type of speech event (activity involving language) might students do asked to do involving this text?**
    - categorize other plants based on the definitions included
    - make arguments for defining plants as high, moderate or limited based on the criteria
- **What grammatical forms are used? Is there a particular tense that is prevalent?**
    - Primarily uses present simple except when describing the process, when the past tense is used
- **What kinds of sentences are used?**
    - Complex sentences are used with dependent clauses and transitional terms
- **Do you see any evidence of a grammatical metaphor (e.g., nominalization—verbs that turn into nouns, e.g., *observe vs observation*)?**
    - Categorizes, categorized
- **What kinds of words and phrases are used?**
    - *General everyday vocab*: plants, land, or, they, etc.
        - Note that more familiar terms like high, moderate, and limited are used in content-specific ways; a word like "they" can also be referring to a challenging concept.

Figure 7. Example of an extended anticipation guide, used for priming comprehension before the reading (left-hand column) and after the reading (right-hand column) through a scaffolding technique for a complex text for adults, modeling the support linguistically diverse students may need for challenging content

Handout 3.A
Module 1
Session 3: Academic Styles of Language in Science

Extended Anticipation Guide

Agree/ Disagree	Statement	Agree/ Disagree	Justification/Page #
	It is important for teachers to know how conversational and academic styles of language are intertwined.		
	The focus in secondary science texts is on observations of the natural world.		
	Conversational styles are developed by adolescence.		
	More students fail or quit school because they can't handle academic language than because they cannot decode text.		
	There are numerous academic styles of language.		
	Some students may see the identity of a scientist and the ways of using language in science as conflicting with their other identities.		
	Academic language styles include spoken language		

Figure 8. Example module activity: Round Robin Writing. In Round Robin Writing, groups of three teaching fellows worked together and were assigned three pages with three separate questions. All three teachers answered the original question as Person A, rotated their papers, and then responded to both the printed question and their colleagues' answers.

Round Robin Writing

- **Get into groups of 3.**
- **You will each get a paper, 1-3.**
- **Each of you will answer each question.**
- **First, start with the question in front of you.**
- **Then switch amongst the three of you.**
  - Read your colleague's response to the question.
  - Comment on your colleague's response and/or with your own ideas.
- **Switch again and repeat the same process.**
- **At the end, pass the papers around and read each response.**

generate conversation and model an egalitarian way to share ideas by prompting more outgoing teaching fellows to wait their turn, while encouraging more passive or shy fellows to participate with plenty of scaffolding. With sentence stems, this activity is an engaging way to get less English proficient K-12 students involved and prevent some of the more proficient students from dominating the discussion.

4. Effective Professional Development Is Well-Organized and Facilitated

Professional development that is well-structured and facilitated is essential to model best practice for educators. Linguistically and culturally diverse learners need routines in order to connect ideas from one session or lesson to another. One way to meet this standard is by sharing and reviewing the goals, agenda, and objectives of the PD session so that they are made clear to participants. Effective PD continuously assesses the needs of participants, and the facilitator adapts to the context of the participants, depending on whether they have more or less background knowledge than anticipated or to tailor them to program models, policy, or student groups. High-quality PD should also include modeling, demonstrations, clear instructions and

Figure 9. Example graphic organizer for the Round Robin Writing activity. By the end of the activity, teaching fellows have shared ideas from two colleagues, modeling an activity to build written interaction.

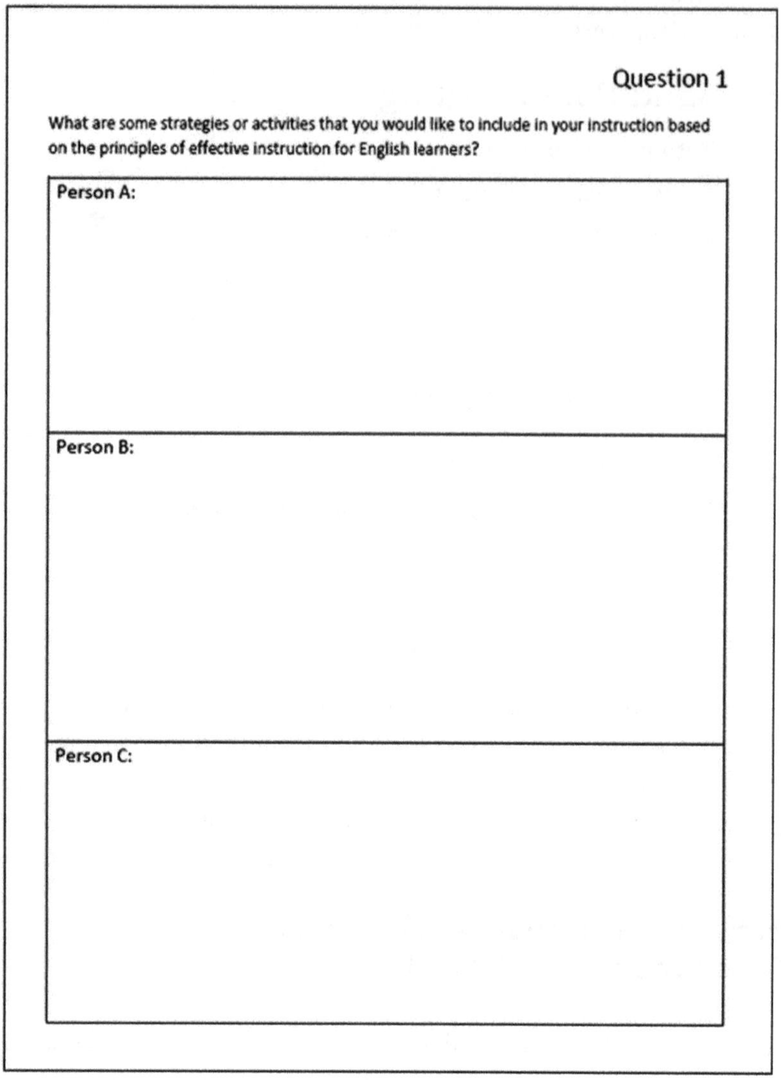
Question 1

What are some strategies or activities that you would like to include in your instruction based on the principles of effective instruction for English learners?

Person A:

Person B:

Person C:

explanations, adequate pacing and time frame, discussions of various perspectives, and well-organized materials.

One key to the success of the language and literacy modules for the MAT program was the organization of materials for both the faculty members as well as for the teaching fellows. The experts were responsible for two modules of material: One

Figure 10. Example module activity: Traveling Jigsaw. In order to build meaningful discussions based on complex journal articles, teachers were divided into jigsaw groups based on different articles and took notes as a team in the four-quadrant graphic organizer, which they then created as a poster. The teaching fellows rotated around the room to read each group's posters and each had an opportunity to present the content to colleagues.

Handout 2.C
Module 1
Session 2: Linguistic Aspects of Academic Language

Summary Points	Implications for Teaching
Examples of Academic Language/Summary Graphic	**Notable Quote or Question for Discussion**

Figure 11. As participants rotated around the room during the Traveling Jigsaw, sample sentence stems were posted to model best practice for linguistically and culturally diverse students.

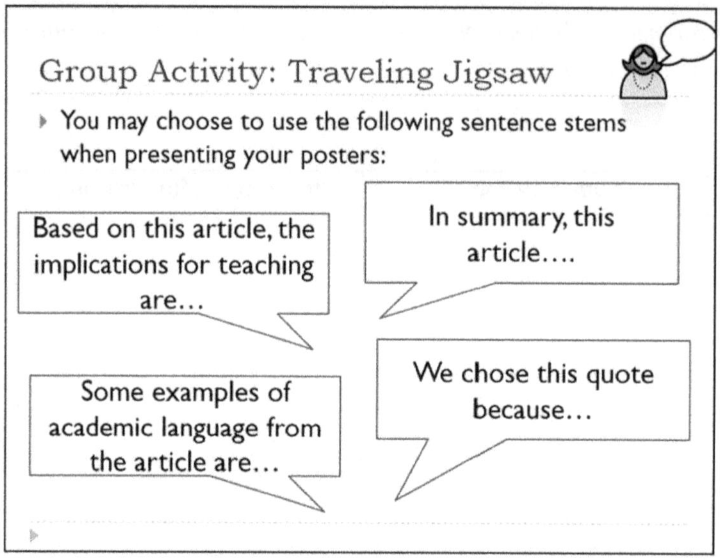

Figure 12. Example module activity: Talking Chips. After reading each quote, participants shared their ideas by placing a talking chip, such as a Lego or paperclip into the center of the table. Each participant had a limited number of chips to control the discussion before a free-flowing conversation can occur.

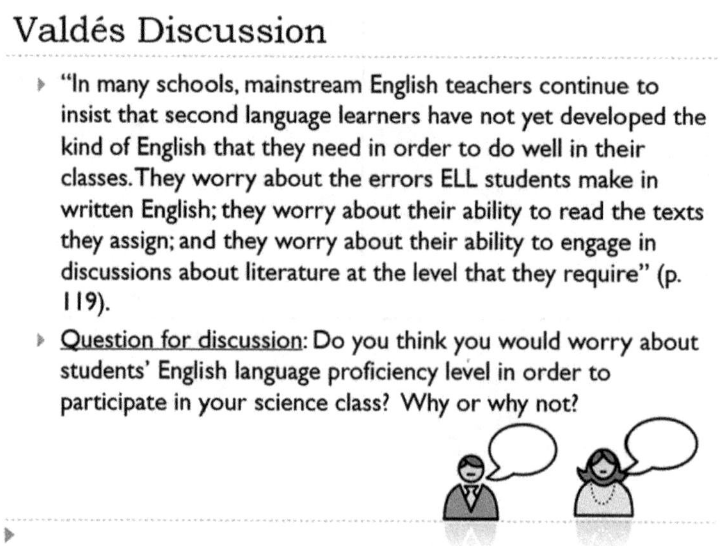

Figure 13. Question or sentence prompts that can be interpreted in different ways during the Talking Chips activity because of ambiguity or personal opinion often generate the most conversation

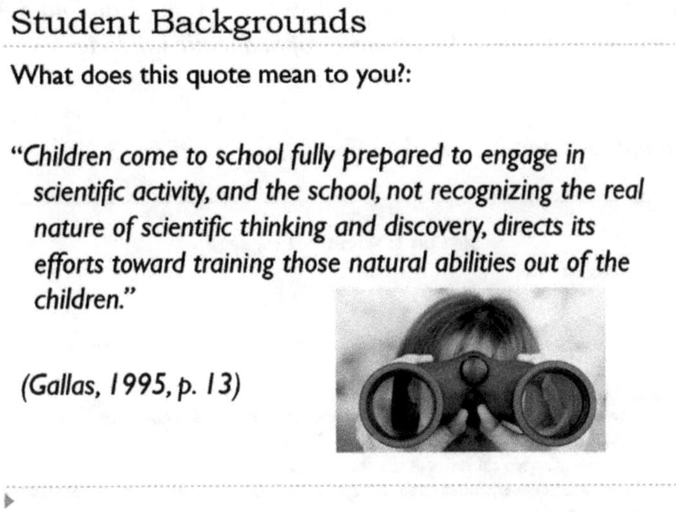

in the fall course, providing content for three class sessions, and one in the winter course, spanning four class sessions. Knowing that the MAT faculty members would not have much time to review, prepare, and duplicate materials, the experts created two-page lesson plans for each classroom session. One page consisted of lesson management information (i.e., suggested pacing, objectives, materials, assigned readings, and any preparation; see Figure 14), and the other consisted of easy-to follow lesson activities (Figure 15). This organized framework also extended to the student materials. Each classroom session consisted of the same lesson components: warm-up; presentation; practice; application; wrap-up; and assignment (see Figure 16). Lessons also began with content and language objectives to model the importance of creating clear lessons for students with measurable outcomes (Figure 17).

The consistency of materials helped the faculty members prepare the course work, determine which sections could be adapted for online work, and gradually progress from reading the full lesson plans to concentrating more on the actual lesson materials: the PowerPoints and handouts. The PowerPoints were scripted with notes under each slide and each slide was recorded so that they could be easily integrated into an online course. A speaker icon in the corner of each slide could be played to access the recording (see Figure 18), giving faculty members the option to use the scripts in person or to have students listen to the entire presentation independently. References were made to each handout, which were labeled according to a standard convention by module and session number followed by the title and whether it was a handout to be provided to fellows or an answer key (e.g., *Handout 3.A_extended anticipation guide; Handout 3.B_examples of academic language*).

Figure 14. Sample session lesson plan

Session 1: Scientific Language

Lesson Management

Pacing

- This lesson is approximately 3 hours of class time. Modifications may be made.

Learning Objectives

- To review and discuss information about the types of language customarily used by scientists and their meaning making potentialities.
- To problematize typical views of scientific language.
- To interview another student about the way language was used in his/ her professional science setting.
- To examine specific uses of scientific language in context.
- To critically think about scientific language practices and implications for pedagogy and student learning.

Materials

- LCD projector
- Handouts
  - Handout 1.A: Ochs review
  - Handout 1.B: Interview Template
  - Handout 1.C: Writing Assignment

Texts

- Transcripts
- Ochs, E., Gonzalez, P., Jacoby, S. (1996). "When I come down, I'm in the domain state": Grammar and graphic representation in the interpretive activity of physicists. In E. Ochs, E. Schegloff, and S. Thompson (Eds.) *Interaction and Grammar*, Cambridge: Cambridge University Press.

Preparation

- Prior to this session, teacher candidates should be instructed to audio record a conversation at their field site (approximately 20-30 minutes). They should listen to the conversation again and choose a small part of the conversation to transcribe (approximately 3-5 minutes) for use in class. This assignment may be given up to two weeks in advance. Prior to this class, tell teacher candidates to bring 4 copies of their transcribed conversation to share with other students. Instruct teacher candidates not to use their field site co-workers' real names and to be sensitive about the information that is transcribed.
- Assign the Ochs, Gonzales, & Jacoby (1996) article. Tell teacher candidates that it is a different kind of article and may be difficult to read. You may choose to give them

Figure 15. Step-by-step directions for faculty, which corresponded to the handouts and slides. Once faculty members were more familiar with the lesson sections, they could refer more to the presentation materials (i.e., PowerPoints and handouts) than to the routine lesson plans.

Lesson Activities

Warm-Up	• Project warm up from power point on screen. • Read warm up questions aloud. • Have teacher candidates write their ideas on a piece a paper. • Ask teacher candidates to talk with a partner about their responses. • Debrief as a whole class. Accept all answers.
Presentation	• Distribute Handout 1.A, "Ochs Review" • Read aloud the slide with review questions. Tell teacher candidates to go through the Ochs article and answer questions as a group. Tell teacher candidates they have 10-15 minutes to complete this activity. It is just a chance for them to review the article together and prepare to discuss it as a class. • When teacher candidates appear to be finished/ almost finished (approx 10-15 minutes), tell them that you will review the answers together. • Review answers using Answer Key 1.A, "Ochs Review". Note that these are possible answers. Go through respective slides and discuss any questions teacher candidates might have.
Practice	• Tell teacher candidates to take out the transcripts from their field placement recordings. • Distribute Handout 1.B, "Interview Activity." • Read the directions aloud. Tell teacher candidates that they have 10 minutes to interview one another about the recording they did at their field site and then they should switch roles (5 minutes each). • Debrief with the whole class (if feasible); if there are too many teacher candidates to debrief with everyone, ask a few teacher candidates about salient themes from their interview.
Application	• Data workshop. Tell teacher candidates to get into small groups of 3 or 4 teacher candidates. They should select one transcript from the group to analyze • Teacher candidates should read the transcript together and characterize the language of the transcript based on their understandings of scientific language (e.g., formal vs. informal, technical vs. every day, detached vs. involved). Their impressions can be general at this point, but they should be attending to the

5. Effective PD Should Be Positively Framed, Respectful and Inclusive

Just as relevant to the instruction of K-12 students, effective professional development should value and integrate participants' backgrounds, experiences, and perspectives into the learning process. PD should promote the inclusion of all stakeholders in diverse learners' success: teachers, families, administrators, specialists, and the

Figure 16. Example overview slide from a session on "Listening and Speaking in Science Learning Contexts"

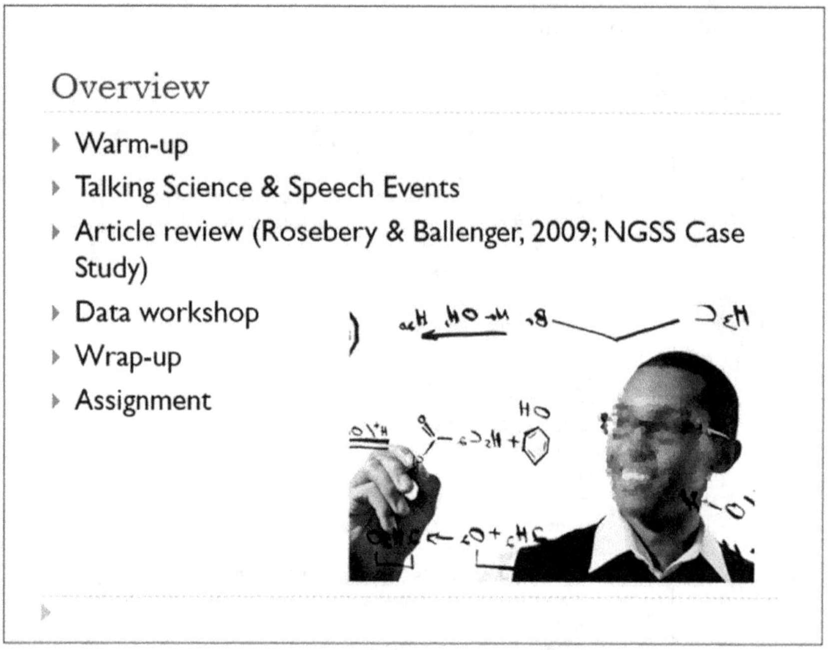

individual student. As a model for ideal K-12 instruction, PD should strive for cultural proficiency, valuing all cultures, languages, and backgrounds as assets in learning, as well as welcoming and respecting participants' diverse opinions and points of view. Based on diverse viewpoints, facilitators should review and consider participant feedback in order to inform and improve future PD sessions. Both PD content and the facilitators should reflect an additive (Cummins, 2000) vision of bilingualism, biliteracy, and biculturalism for students, families, and educators.

Some ways in which the modules and PD sessions for this project met this standard includes an exploration of linguistic variation and understanding of dialect, regionalisms, and accents. Teaching fellows discussed the various biases that exist based on perceptions of a "standard English" and a "monolingual ideology" (Van Lier & Walqui, 2012). Additionally, expert facilitation skills helped to promote mutual understanding during any potentially contentious debates (i.e. restating what the speaker believes in a non-judgmental fashion or asking, "What does everyone else think?"). When teaching fellows seemed to assign differential value to various language styles, the experts added some clarifying activities to a follow-up lesson to ensure that the fellows would adopt an additive viewpoint. The slides also incorporated a variety of pictures of diverse student populations, referred to

Figure 17. Example content objectives slide. Each lesson began with objectives to model the importance of stating learning targets, particularly for linguistically diverse students who need more explicit attention to the content concepts or language learning that they will be expected to know based on the lesson.

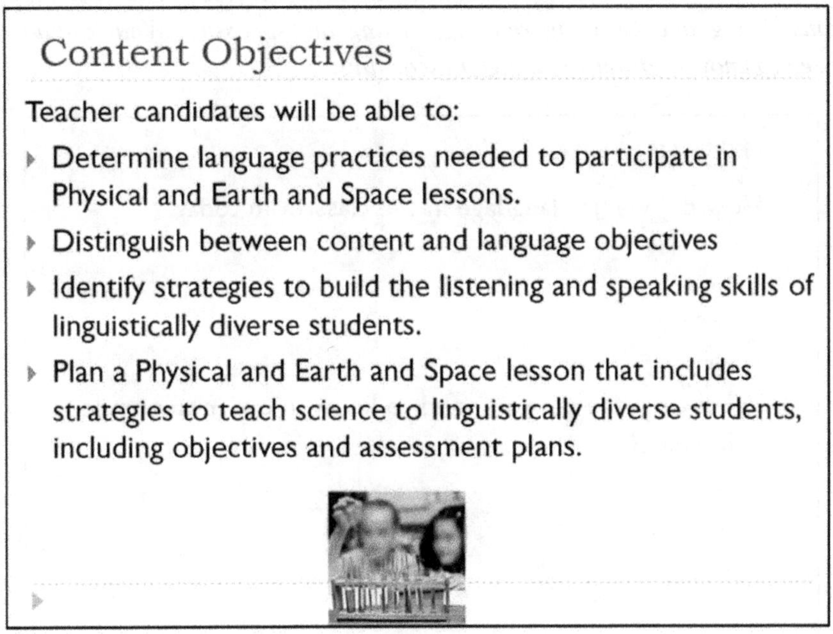

non-dominant language groups, and gave teaching fellows an opportunity to walk in the shoes of a newcomer student by exposing them to a lesson in Mandarin Chinese and asking them to complete a task. One focus of the project was increasing fellows' understanding of language variation, including African American dialect. To support this goal, many of the assigned readings were specifically chosen based on their definition of academic language as a developmental process for all learners and in opposition to the marginalization of ELs (Valdés, 2004).

6. Effective and Sustaining PD Should Be Supportive of Future Learning and Growth

As in all learning focused on retention, PD should support a sustained long-term approach to implementing and reflecting on practices learned in a workshop (Zepeda, 2012). The PD should provide options and ideas for effectively sustaining and growing the knowledge base of the participants. Professional development workshops should be measurable in terms of achieving session objectives, participant satisfaction,

Figure 18. Each slide had a speaker icon to represent the accompanying recorded script. The script appeared as a Note under the slide but was also recorded for professors to listen to in preparation for the class or for the students to listen to as a video presentation during a hybrid class. For this wrap-up slide the script slide was "Think about how we used language in the class today. How did we use the four domains of language: listening, reading, writing, and speaking? What language did you need to know in order to engage in learning?"

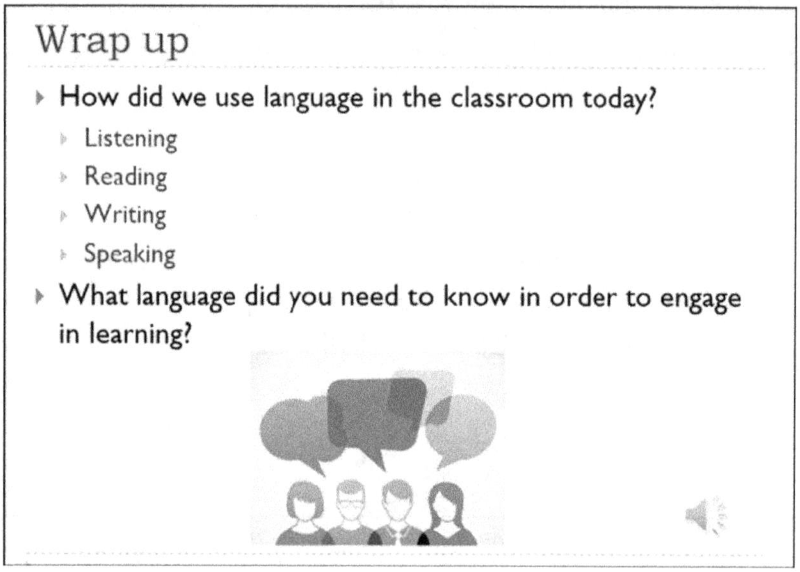

teacher growth, and student outcomes. As in a gradual release model of instruction, the focus is on an individual growth mindset, release of facilitator responsibility, and resources for future learning.

One way that the experts were able to meet these goals was to record the language and literacy modules for use beyond the scope of the project and for viewing outside of class time. The lead MAT faculty member reported downloading the recorded modules as a podcast for her commute. The decision to record the modules was also made with the knowledge that program faculty or the structure of the courses might change over time. The recorded materials, which are archived on a shared drive, can be implemented in other courses and with future cohorts. In order to further build capacity, the experts also co-presented with faculty members during the courses and attended each lecture to answer any specific questions about language acquisition, language instruction, academic language, or other linguistic theory. The experts also delivered academic presentations with both faculty members and teaching fellows at conferences to solidify knowledge and build capacity. During PD sessions for the entire program faculty, the language experts also modeled effective pedagogical techniques and integration of content and language, which through anecdotal reports, sparked some modified teaching approaches. Further project work will include classroom observation of the teaching fellows with coaching to strengthen the practice element of effective professional development.

DISCUSSION AND RECOMMENDATIONS

This project highlights the critical need for including thoughtfully designed content on language in pre-service and in-service STEM educator training. Most of the program participants, including program mentors from participating schools and MAT faculty, struggled with the linguistic content, and many had little to no experience with the notion of academic language. Research shows that this is typical for content-area teachers, for whom linguistics-related courses are rarely required, in spite of the fact that teacher language awareness has been shown to be critical for effective content-area teaching (Lindahl & Watkins, 2015). In addition, science teachers also receive minimal, if any training, to prepare them to work with ELs, even though ELs are included in nearly all classrooms across the U.S. (Lucas, 2011). This gap in teacher training has a profound effect on the educational needs of linguistically and culturally diverse learners. The National Council on Teaching Quality found that in 2014, only 24% of program literacy coursework for elementary educators adequately addressed strategies for ELs (Greenberg, Walsh, & McKee, 2015). According to the Education Commission of the States (2014), more than 30 states do not require EL training for general classroom teachers beyond the federal requirements even though "research indicates that it is beneficial for ELLs if states require all general classroom teachers to have some form of ELL training, regardless of whether they work with ELLs or not." Based on feedback from surveys and "tickets out" from classes and training events, fellows reported that after engaging with the module content they:

- Had a better understanding of the components of academic language;
- Were better able to view the lesson from the student's perspective;
- Were more aware of the challenges faced by linguistically and culturally diverse learners;
- Gained new ideas for activities that could be incorporated into their classroom; and
- Had a better sense of how to modify activities so that they were maximally beneficial to students at all levels of language proficiency.

These shifts in both thinking and practice are vital for science educators, particularly in light of the dual challenges of the linguistic demands of science (Quinn, Lee, & Valdés, 2012) and the growing population of linguistically and culturally diverse students in K-12 schools. English learners and minority students underperform on standardized content assessments compared to their English-speaking peers (NAEP, 2015) and later, linguistically and culturally diverse professionals are underrepresented in science fields, highlighting the need for more robust support for and by educators in the K-12 STEM content classroom. In addition, for those K-12 students who do continue in STEM fields, it is critical that they have the skills to communicate their scientific findings. All STEM students can benefit from learning effective ways to explain observations, make hypotheses, offer counterevidence, or stake claims, very sophisticated linguistic skills.

The response from educators across the project highlight the need for more training in this area. During the project, the experts found that not only were the program fellows interested in learning more about language and linguistics, but the mentor STEM teachers in the partner school district and the university faculty (who were experts in either education or various STEM fields) found the content helpful as well. The research team conducted several professional development workshops for the fellows, mentor teachers, and program faculty, and all participants reported that the information presented was relevant to their classrooms and in many cases made them revisit their own instructional practices from K-12 to university classroom contexts. In fact, there was interest in including training on language and literacy practices in the classroom in college-wide workshops for university faculty outside of the MAT program, given the positive response from the science and education faculty in the program. This reaction highlights both the need for and a strong interest in this content among educators at all levels.

The project also benefited from the collaborative relationship between the linguistics experts, MAT faculty, and teaching fellows, which allowed the materials to be as efficacious as possible. The classroom observations and regular communication with the both the university and district mentor teachers provided many opportunities to modify and adapt content based on student and instructor feedback and program needs. This ensured that the modules were as relevant and meaningful as possible for the program fellows. One of the main modifications to the materials was to augment the content related to linguistic variation. Although this topic was included in an earlier course, the experts noted during one of the subsequent classroom observations that the fellows lacked a strong understanding of linguistic differences and what they might look like in the science classroom. To help solidify participants' understanding of the topic, later modules included a review of this topic as well as additional examples, which seemed to help clarify the content for the cohort. Another example of an adjustment of the language and literacy content to meet program needs was the addition of a lesson on experiencing classroom content in Mandarin from a language learner's perspective. Several of the fellows commented that this activity allowed them to empathize with EL students in their classroom, and it highlighted the critical role that scaffolds and other supports play for students from linguistically diverse backgrounds. Both of these examples show the importance of tailoring training materials to meet program needs, as well as the necessity of providing pre- and in-service teachers with concrete, relevant examples that explicitly connect to their classroom experiences.

Throughout the development and delivery of the language and literacy modules, the experts derived several important lessons from the project experience. First, they found that recording narrations on the PowerPoint slides was an effective way to share the content and ensure that the modules would be useful in the future. While current faculty were trained on the module content, it is likely that over time different instructors, who would not have the benefit of face-to-face training on the content, will be teaching the language and literacy topics in future years. By recording the PowerPoints, the research team provided support to future faculty instructors and ensured that the content was delivered as intended, since the faculty teaching the courses are experts in science education, not linguistics.

Based on feedback from participants across the program, the research team found that the format of the modules worked well. In particular, fellows, mentors, and faculty reported that the modeling of classroom-ready activities throughout the module content was particularly relevant and valuable. On several occasions, program fellows who were in the classroom voiced an intention to use activities in their own classes, and many reported their experiences implementing the activities during future classes. The experts also found that using supplementary PD as a vehicle for delivering the content to program participants was effective. These sessions were outside of the program courses and allowed the experts to develop tailored PD content based on classroom observations in the MAT programs, questions from program fellows, requests from university faculty, and reported needs in the middle school classroom. For the teacher mentors and university faculty, it provided them with training on language and literacy content that is outside their areas of expertise but is (a) a key component of the MAT program content and (b) is relevant to their own diverse student populations. Faculty reported that their awareness of the language demands in their content areas increased and that they used some of the interactive strategies to teach their own content. For the fellows, it was an opportunity to revisit, clarify, and solidify language-oriented topics from the program. For all involved, using the PD approach as a supplement to the modules increased the impact of the linguistic content across the program.

One suggestion that other programs might consider is that while the linguistics experts enjoyed a positive, collaborative relationship with partners across the program, they largely developed the content in isolation. The program would have benefited from integrated planning that included all partners from the start of the program, rather than working independently during the planning process. This might have reduced the need for unplanned adjustments to the program content, like those described earlier. More inclusive planning across the project might also have allowed for the collection of course artifacts, like assignments and lesson plans created by the teaching fellows, that would have supplemented the data collected via evaluations and observations of MAT classes. This data would have provided a more comprehensive assessment of how the teaching fellows were applying the language and literacy content in their own teaching.

CONCLUSION AND FUTURE RESEARCH DIRECTIONS

Future plans for this work aim to increase the efficacy of the language and literacy content by including more classroom-based observations and directed feedback with the program fellows. While linguistics experts observed the implementation of the modules in the university courses, a major question is whether the fellows are accurately and effectively actualizing their knowledge in the classroom, especially with their linguistically and culturally diverse students. To examine this, the linguistics experts plan to observe the fellows in their science classroom contexts with an expert-created protocol. As part of these observations, they will incorporate a feedback cycle that includes individualized coaching to help support the fellows as they increase their understanding of how to apply their knowledge of language and literacy in the science classroom. Coaching as part of a long-term sustained professional development program is an effective practice to involve teachers in a "change process" and allows the work to be tailored to the classroom setting where the observations and feedback can be individualized to the strengths and beliefs of the teachers (Zepeda, 2012). A longer-term goal is to use the information gathered during the classroom observations and from program fellows' work samples and lesson plans, to learn more about the long-term impact of the modules on teaching practices and how this might affect student learning outcomes in the science classroom. Given the incredibly diverse nature of the urban classroom, this work is critical to providing theoretical guidance and practical tools to educators in urban schools where linguistic variation is especially high. Moreover, as linguistically and culturally diverse professionals are underrepresented in STEM fields, this work is particularly significant to preparing the next generation of scientists.

ACKNOWLEDGMENT

This research was supported by the National Science Foundation [Grant #1540799]. The authors would like to thank Dr. Laura Wright for her leadership and expertise in developing the academic language module content. They are also grateful to the program faculty and teaching fellows who worked with the materials, allowed the authors to observe them, and provided feedback on the modules.

REFERENCES

Alfouaim, A. (2012). *The critical language awareness impact on language variation*. California State University.

Alim, S. H. (2005). Critical language awareness in the United States: Revisiting issues and revisiting pedagogies in a resegregated society. *Educational Researcher*, *34*(7), 24–31. doi:10.3102/0013189X034007024

Artiles, A. J., Harry, B., Reschly, D. J., & Chinn, P. C. (2002). Over-identification of students of color in special education: A critical overview. *Multicultural Perspectives*, *4*(1), 3–10. doi:10.1207/S15327892MCP0401_2

August, D., Artzi, L., & Mazrum, J. (2010). *Improving science and vocabulary learning of English language learners (CREATE Brief)*. Washington, DC: Center for Research on the Educational Achievement and Teaching of English Language Learners.

Bishop, R., Berryman, M., Cavanagh, T., & Teddy, L. (2009). Te Kotahitanga: Addressing educational disparities facing Māori students in New Zealand. *Teaching and Teacher Education*, *25*(5), 734–742. doi:10.1016/j.tate.2009.01.009

Bunch, G. C. (2013). Pedagogical language knowledge: Preparing mainstream teachers for English learners in the new standards era. *Review of Research in Education*, *37*(1), 298–341. doi:10.3102/0091732X12461772

Center for Applied Linguistics. (forthcoming). *Standards of professional development excellence for teachers of English learners*. Washington, DC: Author.

Council of Chief State School Officers. (2012). *Framework for English language proficiency development standards corresponding to the Common Core State Standards and the Next Generation Science Standards*. Washington, DC: Author.

Counts, J., Katsiyannis, A., & Whitford, D. K. (2018). Culturally and linguistically diverse learners in special education: English learners. *NASSP Bulletin, 102*(1), 5–21. doi:10.1177/0192636518755945

Cummins, J. (2000). *Language, power, and pedagogy: Bilingual children in the crossfire*. Clevedon, UK: Multilingual Matters. doi:10.21832/9781853596773

Echevarría, J., Richards-Tutor, C., Chinn, V. P., & Ratleff, P. A. (2011). Did they get it? The role of fidelity in teaching English learners. *Journal of Adolescent & Adult Literacy*, *54*(6), 425–434. doi:10.1598/JAAL.54.6.4

Echevarría, J., Vogt, M. E., & Short, D. (2017). *Making content comprehensible to secondary English learners: The SIOP model*. Boston: Allyn & Bacon.

Education Commission of the States. (2014). *50-state comparison: What ELL training, if any, is required of general classroom teachers?* Available at http://ecs.force.com/mbdata/mbquestNB2?rep=ELL1415

Fairclough, N. (1992). *Language and power*. London: Longman.

Galguera, T. (2011). Participant structures as professional learning tasks and the development of pedagogical language knowledge among preservice teachers. *Teacher Education Quarterly*, *38*, 85–106.

Gándara, P., Maxwell-Jolly, J., & Driscoll, A. (2005). *Listening to teachers of English language learners: A survey of California teachers' challenges, experiences, and professional development needs*. Santa Cruz, CA: The Center for the Future of Teaching and Learning.

Gay, G. (2010). *Culturally responsive teaching: Theory, research, and practice* (2nd ed.). New York: Teachers College Press.

Greenberg, J., Walsh, K., & McKee, A. (2014). *NCTQ teacher prep review: A review of the nation's teacher preparation programs*. Washington, DC: National Council for Teacher Quality.

Halliday, M. A. K., & Martin, J. R. (1993). *Writing science: Literacy and discursive power*. Pittsburgh, PA: University of Pittsburgh Press.

Hawkins, J. D., Guo, J., Hill, K. G., Battin-Pearson, S., & Abbott, R. D. (2001). Long-term effects of the Seattle social development intervention on school bonding trajectories. *Applied Developmental Science*, *5*(4), 225–236. doi:10.1207/S1532480XADS0504_04 PMID:17955057

Heath, S. B. (1982). Questioning at home and at school: A comparative study. *Doing the ethnography of schooling: Educational anthropology in action,* 102-131.

Heath, S. B. (1983). *Ways with words: Language, life, and work in communities and classrooms*. Cambridge, UK: Cambridge University Press. doi:10.1017/CBO9780511841057

Hollins, E., & Guzman, M. T. (2005). Research on preparing teachers for diverse populations. In M. Cochran-Smith & K. Zeichner (Eds.), *Studying teacher education: The report of the AERA panel on research and teacher education* (pp. 477–548). Mahwah, NJ: Lawrence Erlbaum.

Hudicourt-Barnes, J. (2003). The use of argumentation in Haitian Creole science classrooms. *Harvard Educational Review, 73*(1), 73–93. doi:10.17763/haer.73.1.hnq801u57400l877

Johnson, C. C., Zhang, D., & Kahle, J. B. (2012). Effective science instruction: Impact on high-stakes assessment performance. *RMLE Online, 35*(9), 1–14. doi:10.1080/19404476.2012.11462092

Kagan, S., Kagan, M., & Kagan, L. (2016). *59 Kagan Structures: Proven Engagement Strategies*. San Clemente, CA: Kagan.

Lee, O., Maerten-Rivera, J., Penfield, R. D., LeRoy, K., & Secada, W. (2008). Science achievement of English language learners in urban elementary schools: Results of a first-year professional development intervention. *Journal of Research in Science Teaching, 45*(1), 31–52. doi:10.1002/tea.20209

Lightbown, P. M., & Spada, N. (2006). *How languages are learned.* Oxford, UK: Oxford University Press.

Lindahl, K., & Watkins, N. M. (2015). Creating a culture of language awareness in content-based contexts. *TESOL Journal, 6*(4), 777–789. doi:10.1002/tesj.223

Lucas, T. (2011). Language, schooling, and the preparation of teachers for linguistic diversity. In T. Lucas (Ed.), *Teacher preparation for linguistically diverse classrooms: A resource for teacher educators* (pp. 3–17). New York, NY: Routledge.

Michaels, S. (2006). Narrative presentations: An oral preparation for literacy with first graders. In J. Cook-Gumperz (Ed.), *The Social Construction of Literacy* (2nd ed.; pp. 94–116). New York: Cambridge University Press. doi:10.1017/CBO9780511617454.006

National Assessment for Educational Progress. (n.d.). U.S. Department of Education, Institute of Education Sciences, National Center for Education Statistics. Available at https://nces.ed.gov/

National Center for Education, Evaluation and Regional Assistance. (2016). U.S. Department of Education, Institute of Education Sciences, National Center for Education Statistics. Available at https://nces.ed.gov/

National Education Association and National Association of School Psychologists. (2007). *Truth in labeling: Disproportionality in special education*. Retrieved from http://www.nea.org/assets/docs/HE/EW-TruthInLabeling.pdf

National Research Council. (2011). *Successful K-12 STEM education: Identifying effective approaches in science, technology, engineering, and mathematics*. Washington, DC: National Academies Press.

National Research Council. (2012). *A framework for K-12 science education: Practices, crosscutting concepts, and core ideas*. Washington, DC: National Academies Press.

Philips, S. U. (1983). *The invisible culture*. New York: Longman.

Quinn, H., Lee, O., & Valdés, G. (2012). *Language demands and opportunities in relation to Next Generation Science Standards for English language learners: What teachers need to know*. Paper presented at the Understanding Language Conference, Stanford, CA. Retrieved from http://ell.stanford.edu/papers

Rickford, J. R. (1999). *African American vernacular English: Features, evolution, educational implications*. Oxford, UK: Blackwell.

Rickford, J. R. (2005). Using the vernacular to teach the standard. In J. D. Ramirez, T. G. Wiley, G. de Klerk, E. Lee, & W. E. Wright (Eds.), *Ebonics: The urban education debate* (2nd ed.; pp. 18–40). Clevedon, UK: Multilingual Matters. doi:10.21832/9781853597985-004

Rogoff, B. (2003). *The cultural nature of human development*. Oxford, UK: Oxford University Press.

Schleppegrell, M. J. (2004). *The language of schooling*. Mahwah, NJ: Lawrence Erlbaum Associates.

Shifrer, D., Muller, C., & Callahan, R. (2011). Disproportionality and learning disabilities: Parsing apart race, socioeconomic status, and language. *Journal of Learning Disabilities*, *44*(3), 246–257. doi:10.1177/0022219410374236 PMID:20587753

Sleeter, C. E. (2008). Preparing White teachers for diverse students. In M. Cochran-Smith, S. Feiman-Nemser, & J. McIntyer (Eds.), *Handbook of research in teacher education: Enduring issues in changing contexts* (3rd ed.; pp. 559–582). New York, NY: Routledge.

Sleeter, C. E., & Owuor, J. (2011). Research on the impact of teacher preparation to teach diverse students: The research we have and the research we need. *Action in Teacher Education*, *33*(5-6), 524–536. doi:10.1080/01626620.2011.627045

Sullivan, A. L. (2011). Disproportionality in special education identification and placement of English language learners. *Exceptional Children*, *77*(3), 317–334. doi:10.1177/001440291107700304

Valdés, G. (2004). Between support and marginalisation: The development of academic language in linguistic minority children. *International Journal of Bilingual Education and Bilingualism*, *7*(2-3), 102–132. doi:10.1080/13670050408667804

Valencia, R. R. (1997). Conceptualizing the notion of deficit thinking. In R. R. Valencia (Ed.), *The evolution of deficit thinking: Educational thought and practice* (pp. 1–12). Bristol, PA: Falmer Press.

van Lier, L., & Walqui, A. (2012). *Language and the Common Core State Standards*. Paper presented at the Understanding Language Conference, Stanford, CA. Retrieved from http://ell.stanford.edu/papers

Waitoller, F. R., Artiles, A. J., & Cheney, D. A. (2010). The miner's canary a review of overrepresentation research and explanations. *The Journal of Special Education*, *44*(1), 29–49. doi:10.1177/0022466908329226

Warren, B., Ballenger, C., Ogonowski, M., Rosebery, A. S., & Hudicourt-Barnes, J. (2001). Rethinking diversity in learning science: The logic of everyday sense-making. *Journal of Research in Science Teaching*, *38*(5), 529–552. doi:10.1002/tea.1017

Wong Fillmore, L., & Snow, C. (2018). What teachers need to know about language. In C. T. Adger, C. E. Snow, & D. Christian (Eds.), *What Teachers Need to Know about Language* (2nd ed.). Bristol, UK: Multilingual Matters.

Zepeda, S. J. (2012). *Professional development: What works*. Larchmont, NY: Eye on Education.

ADDITIONAL READING

Cummins, J. Basic interpersonal communicative skills and cognitive academic language proficiency: Clarifying the distinction. Retrieved from http://iteachilearn.org/cummins/bicscalp.html

Duguay, A. (2012). CREATE: A comprehensive model for instruction of academic language and literacy in the content areas. CREATE Research Brief. Retrieved from http://www.cal.org/create/publications/briefs/comprehensive-model-for-instruction-of-academic-language-and-literacy-in-the-content-areas.html

Gee, J. P. (2008). What is academic language? In A. S. Rosebery & B. Warren (Eds.), *Teaching science to English language learners: Building on students' strengths* (pp. 57–70). Alexandria, VA: National Science Teachers Association.

Martin, J. R. (1993). Literacy in science: Learning to handle text as technology. In M. A. K. Halliday & J. R. Martin (Eds.), *Writing science: Literacy and discursive power* (pp. 166–202). Pittsburgh: University of Pittsburgh Press.

Ochs, E., Gonzalez, P., & Jacoby, S. (1996). "When I come down, I'm in the domain state": Grammar and graphic representation in the interpretive activity of physicists. In E. Ochs, E. Schegloff, & S. Thompson (Eds.), *Interaction and Grammar* (pp. 328–369). Cambridge: Cambridge University Press. doi:10.1017/CBO9780511620874.007

Rosebery, A., & Ballenger, C. (2008). Creating a foundation through student conversation? In A. S. Rosebery & B. Warren (Eds.), Teaching science to English language learners: Building on students' strengths (pp. 1-12). National Science Teachers Association.

Scarcella, R. (2003). *Academic English: A conceptual framework. Technical Report.* Berkeley, CA: University of California Minority Research Institute; Retrieved from http://www.escholarship.org/uc/item/6pd082d4

Wright, L. J. (2008). Writing science and objectification: Selecting, organizing, and decontextualizing knowledge. *Linguistics and Education, 19*(3), 265–293. doi:10.1016/j.linged.2008.06.004

Zwiers, J. (2007). Teacher practices and perspectives for developing academic language. *International Journal of Applied Linguistics, 17*(1), 93–116. doi:10.1111/j.1473-4192.2007.00135.x

KEY TERMS AND DEFINITIONS

Academic Language: The language that helps students engage effectively with academic content.

African American English (AAE): A variety of English spoken by many African Americans in the United States. Also referred to as African American Vernacular English, Black English, and Ebonics.

Critical Language Awareness: The idea that language conventions and language practices are invested with power relations and ideological processes which people are not aware of.

English Learner (EL): A student who is in the process of acquiring English and speaks a language other than English at home. Also knows as emergent bilinguals (EBs) or English language learners (ELLs).

Linguistics: The scientific study of language, its structure, and use.

Pedagogical Modeling: Content and/or materials that provide students with a model of the desired instructional approaches.

Professional Development (PD): Training, formal education, or advanced professional learning intended to help administrators, teachers, and other educators improve their professional knowledge, competence, skill, and effectiveness.

ENDNOTE

1 For the purposes of this paper, “teaching fellows” or “program fellows” refers to the master’s students enrolled in the MAT program; “students” refers to K-12 students, and “teachers” refers to educators who were not enrolled in the program.

Section 3

Strategies and Practices for Improving Urban K–12 STEM Education

Chapter 6
Gendered and Racial Microaggressions in STEM:
Definitions, Consequences, and Strategies Urban Elementary School Professionals Can Use to Combat Them

Amanda J. Rockinson-Szapkiw
University of Memphis, USA

Katherine Wade-Jaimes
University of Memphis, USA

ABSTRACT

Women and people of color remain underrepresented in science, technology, engineering, and mathematics (STEM) workforce and academia. In this chapter, the authors focus on the experience of girls and students of color in urban STEM classrooms through the lens of microaggressions theory. Within this chapter, the authors define macroaggression and discuss the various types (e.g., microinsults, microinvalidations, microassaults). Consequences of microaggressions are discussed and strategies are presented to address microaggressions within the PreK-12 urban school setting.

DOI: 10.4018/978-1-5225-7814-7.ch006

INTRODUCTION

Within urban schools, teacher, administrators, and school counselors' biases, conscious and unconscious, and their language, verbal and nonverbal, can have a profound effect on students. *Microaggressions*, interactions that consciously or unconsciously disparage others based on characteristics or group membership (Pierce *et al.*, 1978; Sue, 2010), are one way biases manifest themselves every day in the urban school environment. Microaggressions may influence students' persistence, achievement, and identity development in STEM by effecting students' beliefs about who STEM professional are and their ability to succeed in STEM fields. Microagressions are especially harmful to students who are underrepresented in STEM Fields (Morley, 2013), including girls and students of color. Thus, discussion and exploration of macroaggressions in the PreK-12 urban environment is warranted. In this chapter, we define macroaggression and discuss types of microaggressions (e.g., *microinsults*, *microinvalidations*, *microassaults)* (Sue et al., 2007), providing examples of each type of microaggressions within the urban school setting. We then discuss the consequences of microaggressions on STEM achievement, persistence, and identity development. Finally, strategies are presented to address microaggressions within the PreK-12 urban school setting, and STEM classrooms in particular.

DEFINITION OF MICROAGGRESSIONS

Microaggressions have become a common topic recently in the literature. With the emergence of this topic in educational contexts, teachers and students are becoming more aware of the presence and effects of microaggressions. However, the concept of microaggressions and their harmful effects are not new. Microaggressions was a term created in 1978 by Pierce as a way to define the phenomenon of negative interactions African Americans experienced (Pierce et al., 1978). While initially developed to describe negative interactions related to race, microaggressions can also be related to gender, religious orientation, socioeconomic status, or other components of identity (Sue, 2010). Microaggressions are most commonly verbal; however, they may also take a nonverbal form. Moreover, microaggressions can include both conscious or unconscious exchanges in which a student or group of students is disparaged based on characteristics or membership (Pierce et al., 1978; Sue, 2010). Microaggressions can also be committed by individuals internal and external to a specific group. Researchers have identified three types of microaggressions that occur whenever an aspect of a student's identity is ridiculed, challenged, invalidated, or dismissed: (1) *microinsults*, (2) *microassaults*, (3) *microinvalidations* (Sue, 2010).

TYPES OF MICROAGGRESSIONS

Microinsults

Sue (2010) defined microinsults, often done unconsciously and subtly, as "characterized by interpersonal or environmental communications that convey stereotypes, rudeness, and insensitivity and that demean a person's identity" (p. 31). For example, an American Indian (AI) student may be asked by a peer, "How did *you* get to be the teacher's assistant this week for our science project?" The underlying message of the peer may be that American Indians are not qualified to have jobs in science. Microinsults can also occur nonverbally. When a White male math teacher fails to acknowledge students of color in the classroom when they raise a hand to respond to a question, the microinsult or message conveyed by the teacher is that the student of color cannot make an important contribution in the math classroom. Another example of a microinsult is when an elementary science teacher says to an African American student, "Can you avoid using Ebonics in this class? It is not scientific." This communicates that an African American student cannot use Ebonics and be a scientist.

Microinsults can be committed by individuals both internal and external to a group and are often more detrimental when committed by someone internal to the group (Nadal et al., 2013; Sue, 2010). For example, Mr. Hollis teaches algebra at an inner-city middle school. When he returns her graded exam, Shelia, a Latina, slumped down in her chair and whispered, "Another C. I guess it's because I am a girl and girls don't get this math stuff." Although it seems counter-intuitive that a student would make a microagression against their own gender and ethnic group, internalized microaggressions are not uncommon (Kohli and Solórzano, 2012) and demonstrate the deep, lasting damage caused by exposure to microaggressions over time. By promoting the negative stereotype that girls, specifically Latina girls, cannot be successful in math, Shelia is invalidating not only her own ability and identity but also that of other Latina students in the classroom.

Microassaults

Microassaults are less subtle than microinsults and are "conscious, deliberate and either subtle or explicit… biased attitudes, beliefs, or behaviors that are communicated to marginalized groups through environmental cues, verbalizations, or behaviors," (Sue, 2010, p. 28). A microassault can take the form of name-calling, racial epithets (e.g., "colored"), or purposeful discriminatory actions (e.g., deliberately calling on a White male in the classroom before a Latina girl or displaying a picture of only white male scientists in a lab). Diamond's experience provides an example of a microassult. Diamond is an African American high school student from a low socioeconomic (SES) background. Experiencing success with her lab assignment, Diamond noted, "I hope to go to college and pursue a degree in biology." To which, Mrs. Peyton, her teacher, responded, "Oh, honey that is a great aspiration, but it's unreasonable to think that a black girl from this area can ever be a successful scientist." Mrs. Peyton's words communicate to Diamond that being a Black female from a low SES background cannot be a scientist. This influences Diamond's view of science and scientists.

Microinvalidations

Finally, microinvalidations are "characterized by communications or environmental cues that exclude, negate, or nullify the thoughts, feelings, or experiential realities of certain groups" (Sue, 2010, p. 37). Microinvalidation can be conscious or unconscious; however, they are most often unconscious. Helms (1992) stated that African Americans students experience microinvalidations when teachers say "I don't see color" as the comment negates their experiences as racial beings. Mr. Tate's, a school counselor, comment provides another example of a microinvalidation. During a career guidance lesson, Mr. Tate noted that he had heard rumors that the eighth-grade girls did not feel they received equal support and attention from teachers as the boys did during several technology classes. He asserted that all teachers go through extensive training about biases in STEM, so gender bias should not be an issue in STEM courses at the school. He purported that the girls need to be more assertive about their needs. Thus, the career guidance lesson for the day would focus on being assertive. His comments shut down all communication about gender in STEM. The message is sent that the school counselor and the school do not take the girls' concerns seriously and are not open to discussing gender bias in STEM. The communication invalidates the feelings and thoughts of the girls.

WHY MICROAGGRESSIONS IS AN IMPORTANT TOPIC IN STEM

So, why do microaggressions matter? Who cares if a teacher or peer assaults, insults, or invalidates an aspect of personal identity? If a student has a strong personal identity and self-confidence, can they simply just ignore the ridicule or dismissal? If not, what psychological and emotional harm or adverse effects of microaggressions are experienced in STEM classrooms within urban settings?

While research on the topic of microaggressions in educational contexts is only emerging, some research has documented the negative effects, primarily in higher education with some focus on PreK-12. Whether a teacher, peer, or other individual employs a micro assault and rationalizes its effect, commits an unknowing microinsult, or engages in microinvalidations, thinking there is no reason for concern, all types of microaggressions cause harm. Negative effects documented in the literature include emotional and cognitive strain. The mental and emotional energy it takes to evaluate a microaggressions encountered can take away from a students' engagement in the classroom and ability to cognitively process (Bair & Steele, 2010), effecting academic achievement. When a microaggressions is committed by an authority figure such as a teacher, school counselor, or administrator, a student may experience psychological stress, which results in decreased performance (e.g., poor grades in a STEM course) and even decreased self efficacy or belief in the ability to achieve (e.g., ability to perform well in STEM) (Mercer et al., 2011; Torres &Driscoll, 2010). Moreover, microaggressions can result in feelings of isolation. If a girl feels isolated in a science class, for example, she may develop anxiety and depression (Nadal et al., 2011). Feeling isolated, she may come to believe that science is unwelcoming to women and decide to not pursue a degree or career in science, even given aspirations she once had to do so. An association between microaggressions and the stereotype threat, fear about conforming to a negative stereotype associate with one's social group, have also been documented (Steele & Arronson,1995). When a microaggressions is encountered, and stereotype threat occurs, a student may find themselves in a STEM class where he or she wants to succeed. However, due to the emotional and cognitive stress, the student may find that he or she is unable to perform optimally. He or she may even underperform to the point of failing, which in turn may limit future domains of study, lead to devaluing the subject, and narrowing degree and career options (Aronson, Fried, & Good 2002; Osborne, 1995; Steele, 1997). When it comes to STEM education, students of color and girls are extremely susceptible to experiencing microaggressions, given the STEM culture. As such, in STEM education, experiencing microaggressions may make it challenging to develop an identity as a scientist or other STEM professional, achieve high grades, or pursue a degree or a career in STEM.

The Culture of STEM

Unfortunately, researchers have documented these various microaggressions within STEM education, including, but not limited to, teachers (1) treating students of color as criminals or second class citizens (e.g., a math teacher only recommending 3% of Latina students for the advanced Algebra class when Latina make up 20% of the schools population), (2) underestimating personal ability by having low expectations for students of color or girls, and (3) making subtle comments or actions based on race (Harrison & Tanner, 2018; Solorzano, 1998; Torres, Driscoll, & Burrow, 2010). Microaggressions documented within the STEM classroom have been verbal, nonverbal, or environmental (Harrison & Tanner, 2018); microaggressions can be consequential of stereotypes and biases.

McGee (2016) argued that microaggressions can be seen as the result of stereotypes about who can participate in STEM fields; i.e., stereotypes that girls and students of color cannot be scientists can create implicit biases in other students and teachers and lead to microinsults, microassaults, and microinvalidations. These stereotypes stem from traditional views of the culture of science and engineering, which has been described as a "chilly" environment for women and people of color (Blickenstaff, 2005). Although it's often described as objective or neutral, researchers have recognized that STEM fields can be viewed as distinct cultures (Cobern & Aikenhead, 1997; Maddock, 1981). Aikenhead (1996) described, for example, the scientific culture as a well-defined system of meanings and symbols that became institutionalized in Europe in the 17th century as a predominantly white, male, middle-class system, referred to today as Western Modern Science (WMS). The cultural features of WMS are described as "mechanistic, materialistic, masculine, reductionistic, mathematically idealized, pragmatic, empirical, exploitive, elitist, ideological, inquisitive, objective, impersonal, rational, universal, decontextualized, communal, violent, value-free, and embracing disinterestedness, suspension of belief, and parsimony" (Aikenhead, 1996, p. 9). Aikenhead argued that the goal of traditional science instruction is the cultural assimilation of WMS, which leads to a culture of school science that perpetuates stereotypic images of science as "socially sterile, authoritarian, non-humanistic, positivistic, and absolute truth" (1996, p. 9). Lemke (1990), and later Cheryan Master and Meltzoff (2015), also argued that schools perpetuate stereotypes about science and other STEM fields as being authoritative, elite, and difficult and portray scientists as being superhuman. This leads to STEM being divorced from everyday life and devoid of emotions, values, or humor. The discourse of STEM in schools is defined by white, male, middle class, native English speakers who are committed to the values of middle-class culture, which includes emotional control, orderliness, rationalism, achievement, punctuality, and social hierarchy.

The culture of WMS and other STEM disciplines are often foreign to students and conflict with their other subcultures, such as family or peer cultures. Therefore, all students must cross borders when they enter STEM classrooms (Aikenhead, 1996). However, it is easier for students who are already familiar with the discourses of WMS and STEM (i.e., white, male, middle-class students) to make this crossing. Meyer and Crawford (2011) adopted the analogy of border crossing, referring to learning science as cross-cultural education that is "laden with cultural understandings, interpretations, and a language of its own" and describing the science classroom as a "borderland of cultural interaction" (p. 531).

Implicit within this culture of STEM is the message that STEM is not for everyone and that, specifically, it is not for girls or students of color. The type of learning environment created by traditional views of the culture of science can lend itself to microaggressions, particularly microinvalidations through subtle messages about who can and cannot perform in STEM, leading to a persistent undervaluing of students of color and girls in STEM (McGee, 2016). Additionally, despite both populations having high interest in STEM during early grades in school, both women and students of color are less likely than white males to pursue science or engineering and other STEM fields as a career, indicating that because the culture of STEM is foreign or inhospitable to them, it is difficult to participate in it fully and develop a scientific identity. Harrison and Tanner (2018) contended that the role of microaggressions is an important, but underappreciated, factor in this underrepresentation.

The Consequences of Microaggressions in the STEM Classroom

As young as 10, students hold gendered views of science. Although all students often enjoy "doing" science, girls are less likely to see science as a potential career, i.e., less likely to see themselves "being" a scientist (Archer et al., 2010, 2012). Archer et al. (2012) found that science identities more often aligned to typical masculine identities, such as being dangerous, and were in tension with typical feminine identities. Both boys and girls state that girls would not make good scientists. Additionally, students see STEM classes as hard, only for naturally smart people, and uncool, which further constrains the science identities. Girls who are interested in STEM are described (by themselves and their parents) as geeky, non-girly, and although not one-sided, interested primarily in academic pursuits (Archer et al., 2012). Ong, Wright, Espinosa, and Orfield (2011) claimed that these stereotypes about women and girls in STEM fields lead to beliefs that women, and particularly women of color, don't belong in science. As an example, she points to a speech given by a Harvard University president, in which he suggested that women might not be willing to work the demanding hours required of elite scientists, or may not have the aptitude for science (Ong, 2005), both forms of microaggressions toward women in science.

Investigating students of color, Archer, deWitt, and Osborne (2015) demonstrated that Black students and parents in the UK are less likely than white students and parents to consider STEM-related careers as "thinkable" options, indicating that students are not forming science identities. Additionally, Black students and parents had narrow, stereotypical conceptions of scientists as wearing white lab coats and goggles, images that were incompatible with students' views of themselves and Black parents and students were more likely to see science as particularly difficult and only for very intelligent, sometimes geeky, students. Black parents were also more likely to see science as a white, masculine undertaking, describing daughters as "girly girls" who were not as likely to be interested in science. Researchers have also shown that African American students experience cultural conflict in STEM classrooms, recognizing that a stigma and cost associated with engaging in STEM (Brown, 2004). This conflict can interfere with learning, which is particularly important to consider as urban classrooms become increasingly diverse and the teaching workforce remains predominantly white (Norman, Ault, Bentz, & Meskimen, 2001). This can lead to significant differences in experiences between white students and students of color science classrooms, including lower expectations from teachers and ignored requests for help (Norman et al., 2001), both forms of microinvalidations.

There is also a connection between microaggressions in STEM communities and a sense of belonging for students and scientists of color. Brown, Henderson, Gray, Donovan, and Sullivan (2013) demonstrated that STEM-related community excludes students of color and creates a cultural mismatch through the use of microaggressions. Moreover, the more microaggressions experienced, the less likely a person of color was to feel aligned with the science community. These microaggressions send messages of exclusions as well as saying that Black students do not have as much to offer the STEM communities as their white peers, that they are less talented and less qualified (Brown et al., 2016). Those students of color who were successful in STEM classes had to tolerate a subtext of racial bias in the scientific community and negotiate expectations from others that were based on racist stereotypes (Brown et al., 2016). Additionally, successful students of color were seen as exceptions to the Black community and subjected to racial tokenism and isolation that exasperated the stress they felt.

Related, researchers have shown that microaggressions can result in isolation and making individuals lack a sense of belonging in the STEM disciplines (Dortch & Patel, 2017; Torres & Driscoll, 2010). Lack of belonging or feeling like an outsider can have negative mental health outcomes, such as depression or anxiety (Donovan et al., 2014; Huynh, 2012; Wang et al., 2011). Learning can be impeded, and underrepresented students may fail to reach, or even recognize, their potential in STEM (Lester, Yamanaka & Struthers, 2016; Torres & Driscoll, 2010) because experiencing microaggressions can lead to the belief that it is impossible to succeed in STEM due to identification with a certain identity or group. Cognitive effort better spent on learning and stress can be a consequence of uncertainty about how to address the microaggressions. Should the student simply endure it or confront it? What if the microaggressions comes from a person in power such as a teacher or principal? Moreover, microaggressions, especially when experienced frequently, can create a hostile culture for students and groups, which can result in underrepresented students deciding to not pursue STEM degrees and careers (Seymour & Hewitt, 1994).

STRATEGIES TO OVERCOME MICROAGGRESSIONS IN THE URBAN SCHOOL SETTING

Although most of the research on microaggressions in STEM has focused on post-secondary education or career paths in science and engineering, microaggressions impact students much earlier, throughout their K-12 education and other STEM experiences. Many of the factors that have been identified as contributing to the alienation of girls and students of color from STEM at young ages can be understood as microaggressions (e.g., Norman et al., 2001). This chapter argues that preventing and addressing microaggressions in K-12 urban classrooms could potentially improve interest in STEM, increase STEM achievement, and support STEM identity work for girls and students of color before they decide not to pursue STEM pathways. Additionally, instead of helping students cope with and overcome microaggressions, we seek to provide recommendations for teachers and other school personnel to proactively combat and prevent microaggressions in urban STEM classrooms. This is especially important for younger students, who may not have the vocabulary and understanding to name their experiences as "microaggressions." These students are therefore at risk of internalizing the negative messages conveyed through microinsults and microinvalidations (Kohli and Solórzano, 2012). The suggestions here focus on three areas: general approaches to addressing microaggressions in the moment; minimizing the impact and occurrence of microaggressions by supporting students' cultural identities in STEM; and creating a STEM learning environment that allows all students to feel comfortable and supported.

GENERAL APPROACHES TO ADDRESSING MICROAGGRESSIONS IN EDUCATION

It is important to acknowledge and address microaggressions when they occur. School personnel need to be able to recognize microaggressions from themselves or others and not downplay or ignore incidents. Here are some strategies for urban schools, teacher, administrators, and school counselors:

Acknowledge and Validate

School personnel need to acknowledge and communicate to students that microaggressions occur and can cause harm. A principal may show a short video clip of microaggressions at a school rally and encourage students to identify and talk to teachers if they see microaggressions. A school counselor could prepare and implement guidance lessons on the topic of microaggressions. Acknowledging microaggressions communicates to students the school personnel's willingness to address microaggressions. Related to acknowledging is validating. When school personnel discusses microaggressions, they need to validate the thoughts and feelings that can be consequential of them, for this communicates further support of those affected.

Talk About It

If school personnel observes or a student shares an experience about microaggressions, the microaggressions should be explicitly discussed, not avoided. Boysen (2012), although drawing from a predominately white population, suggested that students believe that one of the best ways to address racial microaggressions in a classroom environment is a class discussion. Drawing from focus groups with people of color, Sue et al. (2009) concurred that acknowledging and seriously discussing microaggressions with all students in a safe environment is vital to addressing them. Moreover, in these discussions, Sue et al. (2009) emphasized having a person of power (e.g., school counselor, teacher, or principal) guide the discussion is important. For example, if a student brings microaggressions among students to a teacher's attention, the teacher should discuss the incident with the student, talking in private to the student about her thought and feelings. How and why the microaggressions is concerning to the student should be explored. Talk to the student in private about her feelings and why the comment or behavior was troubling. Then, if the student feels safe, the teacher, perhaps with the assistance of the school counselor, should take time to facilitate a conversation among all those involved, deeply exploring the origin of the microaggressions and the negative messages and consequences of them.

Compassionately Confront

Compassionate confrontation of the individual from whom the microaggressions originated is key to eliminating microaggressions in the school environment. School personnel can privately discuss the microaggressions with the individual, remembering that microaggressions are both conscious and unconscious, and use the discussion as an opportunity to educate the offender about the harm of their microaggressions. This is equally important when the microaggressions comes from a member internal to the oppressed group as it is when addressing a microagression from someone external to that group. Drawing from the examples provided above for microinsults, Mr. Hollis could schedule an after-school meeting with Shelia to discuss her microinsult, noting that he perceived Shelia's comment in class as a microinsult promoting the negative stereotype that women and, specifically Latina women, cannot be successful in math. Mr. Hollis can ask Shelia why she feels that way, provide support and encouragement for her to counteract the microaggressions, and may even use the private meeting as an opportunity to explore with Shelia mentoring and tutoring opportunities to help her combat stereotypes she may have adopted related to her gender and racial identity.

RESULTS OF ADDRESSING MICROAGGRESSIONS

Microaggressions can be perpetuated by students or teachers, in either verbal or non-verbal forms, and teachers need to be prepared to explicitly address them. This can be done by explicitly acknowledging a microaggression has occurred, validating students' feelings, and leading discussions either with individual students or in larger groups. Research has acknowledged the importance of teachers explicitly addressing microaggressions in K-12 classrooms. In particular, students have indicated that having a teacher who is comfortable with discussions around race and able to validate the feelings of students of color lead to productive conversations in response to microaggressions that students viewed as valuable learning experiences (Sue et al., 2009).

Supporting Cultural Identities in Science

Brown et al. (2016) found that African-American students who had a strong foundation in African-American history were able to use that foundation as a source of alignment within STEM communities. They argued that in learning environments, supporting students' cultural identities should increase alignment with STEM and decrease the occurrence of microaggressions. This section provides some practical ideas for teachers and other school personnel to support cultural identities of students of color and girls in STEM classrooms; however, it is not meant to be an exhaustive list. It is important to note also that school personnel themselves should be well-versed in the cultural histories and identities of their students in order to support them in the classroom. Without a deep understanding of different cultures, the suggestions here may be perceived as shallow or superficial by students. The suggestions here are presented as a series of questions that teachers can use to reflect on their practice.

Representation of Science

It is important to consider what messages are conveyed to students about STEM disciplines is and who can do STEM. Broad, inclusive views of STEM can help minimize microaggressions in urban classrooms. Some questions teachers and school personnel can reflect on include:

- Do you use examples from different cultures and backgrounds in class? For example, what videos, news articles, or other examples do you use in class? Do they represent a wide range of cultures and backgrounds?
- How is STEM represented in the classroom? When you talk about specific scientists, engineers, or other STEM professionals, what examples do you use? What mentors, guest speakers, etc. are invited into the classroom? When you show a video do you make an effort to find a video featuring women or scientists of color?
- How is STEM represented in clubs or other informal spaces? Does the membership of these groups represent the demographics of the school? If not, have you invited a diverse group of students to join or considered starting a club just for girls or students of color?

History of Science

Understanding one's cultural identity includes understanding one's cultural history and, specifically, the contributions one's culture has made to STEM. Unfortunately, many textbooks and other curricular materials include very little information on the history of STEM or on non-Western contributions to STEM or contributions from women or scientists of color. It's important to acknowledge and address this lack with your students. Some questions to reflect on:

- Consider the textbook or other materials you use regularly. Does it present a comprehensive and accurate picture of the nature of STEM? Does it include non-western contributions to STEM? Does it include contributions from women and scientists of color?
- Do you talk about the history of STEM with your students? In particular, have you discussed why we do not often see women or scientists of colors in textbooks?
- Do you ever talk about or study SETM from other cultures, particularly non-western contributions to science? Consider having students read about or research non-western science.

Creating a Supportive Science Learning Environment

Although there has been much research published on STEM learning environments, it often focuses on logistical concerns like classroom management or learning styles. However, considering the "chilly" culture of science, science learning environments can become sites of cultural conflict for students (Norman et al., 2001) which increase the occurrence of microaggressions. To fully combat microaggressions, teachers and other personnel need to go beyond stopping racist comments or acts (microassaults) to creating an environment and culture that is respectful and supportive of students from all backgrounds. Brown et al. (2016) found that STEM students had various levels of engagement within a community and that to truly belong to the community, they must be aligned to the norms of that community, which decreases the frequency of microaggressions. Therefore, it's important to create an environment that welcomes all backgrounds with norms that do not conflict with students' cultural norms (Brown, 2016). This section focuses on creating an inclusive learning environment, which can minimize the occurrence and impact of microaggressions. Many of these suggestions are based on work around creating equitable learning environments, with a focus on minimizing environmental microaggressions. As with the previous section, the suggestions here are presented as a series of questions for teachers to reflect on their practice.

Classroom Interactions

Daily interaction between teachers and students as well as among students can either create a supportive environment or an environment in which microaggressions flourish. As you consider the interactions in your classroom, below are some questions to reflect on. It may be helpful to video record yourself teaching, or have another teacher observe your class so that you can more accurately reflect on these questions.

- During whole-class discussions, do all students engage in the discussion? When you call on individual students, do you call on a variety of students or the same few over and over? Do the students you call on represent the demographics of the class? Do you have a variety of interactions with each student – i.e., low-level questions, high-level questions, thoughtful dialogue?
- Do you use proximity with students equitably? When students are doing independent or group work, do you visit all students/groups regularly? How would you describe your interactions with each group – i.e., asking questions, clarifying directions, redirecting behavior?

- Do you know the names of all of your students and can you pronounce them correctly?
- When considering recommendations for special opportunities in science, i.e., summer camps, programs, and scholarships, are you recommending a diverse range of students? Are you seeking out and passing along opportunities specifically for girls or students of color?
- Pay attention to non-content language – what are you talking to students about? Does it represent the variety of their interests and cultures? Which students are you talking to?
- When students work together, how do they interact with each other? Do students avoid working or talking with students who are not like them? Are groups heterogeneous by race and gender?

Classroom Policies

Classroom policies can also send an implicit measure to students (i.e., microinvalidation) about who is a valuable member of the science learning community. As you consider your classroom policies and procedures, here are some questions to guide reflection:

- Do you have clear, non-subjective policies around regular classroom occurrences such as going to the restroom or sharpening your pencil?
- If you have class roles or jobs, do students experience a variety of responsibilities or are they limited to stereotypical roles? For example, are the girls in "helper" roles such as passing out papers while the boys are doing more active jobs? This is also an important consideration during labs or other group work when girls are often given the task of taking notes while boys perform the experiment.
- When considering classroom discipline policies and procedures, are they clear and objective? Reviewing your recent disciplinary actions, are they equitably distributed across demographics or are some groups of students more likely to receive punishments?

CONCLUSION

Microaggressions and the various types of microaggressions (e.g., *microinsults, microinvalidations, microassaults)* (Sue et al., 2007) exist within the urban school setting. Dire consequences of microaggressions include, but are not liked to poor STEM achievement, lack of persistence, and identity development. Although much research has focused on coping mechanisms for dealing with microaggressions for college students and professionals, this chapter argues that teachers and other school personnel are in a unique position to address microaggressions earlier in K-12 education. As Kohli and Solórzano (2012) argue, we acknowledge that teachers are in a unique position where they can shape students' perceptions of STEM and create learning environments that are supportive of all students. STEM teachers in particular have an opportunity to choose to perpetuate gendered and racially biased stereotypes of who can do STEM, or to challenge these stereotypes and the norms of STEM and STEM education. The strategies presented here are meant to serve as tools for teachers to address microaggressions within the preK-12 urban school setting, and STEM classrooms in particular, as well as to reflect on their own practice and classroom environment. We hope these suggestions will support both teachers and students in actively addressing and minimizing the impact of microaggressions in K-12 STEM classrooms.

NOTE

This research received no specific grant from any funding agency in the public, commercial, or not-for-profit sectors.

REFERENCES

Aikenhead, G. S. (1996). Science education: Border crossing into the subculture of science. *Studies in Science Education*, *27*(1), 1–52. doi:10.1080/03057269608560077

Archer, L., Dewitt, J., & Osborne, J. (2015). Is Science for Us? Black Students' and Parents' Views of Science and Science Careers. *Science Education*, *99*(2), 199–237. doi:10.1002ce.21146 PMID:28579645

Archer, L., Dewitt, J., Osborne, J., Dillon, J., Willis, B., & Wong, B. (2010). "Doing" science versus "being" a scientist: Examining 10/11_year_old schoolchildren's constructions of science through the lens of identity. *Science Education*, *94*(4), 617–639. doi:10.1002ce.20399

Archer, L., DeWitt, J., Osborne, J., Dillon, J., Willis, B., & Wong, B. (2012). "Balancing acts": Elementary school girls' negotiations of femininity, achievement, and science. *Science Education*, *96*(6), 967–989. doi:10.1002ce.21031

Bair, A. N., & Steele, J. R. (2010). Examining the consequences of exposure to racism for the executive functioning of Black students. *Journal of Experimental Social Psychology*, *46*(1), 127–132. doi:10.1016/j.jesp.2009.08.016

Boysen, G. (2012). Teacher and student perceptions of microaggressions in college classrooms. *College Teaching*, *60*(3), 122–129. doi:10.1080/87567555.2012.654831

Brown, B. A. (2004). Discursive identity: Assimilation into the culture of science and its implications for minority students. *Journal of Research in Science Teaching*, *41*(8), 810–834. doi:10.1002/tea.20228

Brown, B. A., Henderson, J. B., Gray, S., Donovan, B., & Sullivan, S. (2013). From Access to Success: Identity Contingencies & African-American Pathways to Science. *Higher Education Studies*, *3*(1). doi:10.5539/hes.v3n1p1

Brown, B. A., Henderson, J. B., Gray, S., Donovan, B., Sullivan, S., Patterson, A., & Waggstaff, W. (2016). From description to explanation: An empirical exploration of the African-American pipeline problem in STEM. *Journal of Research in Science Teaching*, *53*(1), 146–177. doi:10.1002/tea.21249

Cheryan, S., Master, A., & Meltzoff, A. N. (2015). Cultural stereotypes as gatekeepers: Increasing girls' interest in computer science and engineering by diversifying stereotypes. *Frontiers in Psychology*, *6*, 49. doi:10.3389/fpsyg.2015.00049 PMID:25717308

Cobern, W., & Aikenhead, G. (1997). *Scientific Literacy and Cultural Studies Project*. Scientific Literacy and Cultural Studies Project. Paper 13.

DeWitt, J., Archer, L., & Osborne, J. (2012). Nerdy, Brainy and Normal: Children's and Parents' Constructions of Those Who Are Highly Engaged with Science. *Research in Science Education, 43*(4), 1455–1476. doi:10.100711165-012-9315-0

Dortch, D., & Patel, C. (2017). Black undergraduate women and their sense of belonging in STEM at predominantly white institutions. *NASPA Journal About Women in Higher Education, 10*(2), 202–215. doi:10.1080/19407882.2017.1331854

Harrison, C., & Tanner, K. D. (2018). Language Matters: Considering Microaggressions in Science. *CBE Life Sciences Education, 17*(1), fe4. doi:10.1187/cbe.18-01-0011 PMID:29496676

Helms, J. E. (1992). *A race is a nice thing to have: A guide to being a white person or understanding the white persons in your life*. Topeka, KS: Content Communications.

Huynh, V. H. (2012). Ethnic microaggressions and the depressive and somatic symptoms of Latino and Asian American adolescents. *Journal of Youth and Adolescence, 41*(7), 831–846. doi:10.100710964-012-9756-9 PMID:22453294

Kohli, R., & Solórzano, D. G. (2012). Teachers, please learn our names!: Racial microagressions and the K-12 classroom. *Race, Ethnicity and Education, 15*(4), 441–462. doi:10.1080/13613324.2012.674026

Lemke, J. L. (1990). *Talking Science: Language, Learning, and Values*. Norwood, NJ: Ablex Publishing Corporation.

Lester, J., Yamanaka, A., & Struthers, B. (2016). Gender microaggressions and learning environments: The role of physical space in teaching pedagogy and communication. *Community College Journal of Research and Practice, 49*(11), 909–926. doi:10.1080/10668926.2015.1133333

Maddock, M. N. (1981). Science education: An anthropological viewpoint. *Studies in Science Education, 8*, 1–26.

McGee, E. O. (2016). Devalued Black and Latino Racial Identities. *American Educational Research Journal, 53*(6), 1626–1662. doi:10.3102/0002831216676572

Mercer, S. H., Zeigler-Hill, V., Wallace, M., & Hayes, D. M. (2011). Development and initial validation of the Inventory of Microaggressions Against Black Individuals. *Journal of Counseling Psychology*, *58*(4), 457–469. doi:10.1037/a0024937 PMID:21842986

Meyer, X., & Crawford, B. A. (2011). Teaching science as a cultural way of knowing: Merging authentic inquiry, nature of science, and multicultural strategies. *Cultural Studies of Science Education*, *6*(3), 525–547. doi:10.100711422-011-9318-6

Morley, L. (2013). The rules of the game: Women and the leaderist turn in higher education. *Gender and Education*, *25*(1), 116–131. doi:10.1080/09540253.2012.740888

Nadal, K. L., Issa, M., Leon, J., Meterko, V., Wideman, M., & Wong, Y. (2011). Sexual orientation microaggressions: "death by a thousand cuts" for lesbian, gay, and bisexual youth. *Journal of LGBT Youth*, *8*(3), 234–259. doi:10.1080/19361653.2011.584204

Nadal, K. L., Sriken, J., Davidoff, K. C., Wong, Y., & McLean, K. (2013). Microaggressions within families: Experiences of multiracial people. *Family Relations*, *62*(1), 190–201. doi:10.1111/j.1741-3729.2012.00752.x

Nadal, K. L., Wong, Y., Griffin, K. E., Davidoff, K., & Sriken, J. (2014). The adverse impact of racial microaggressions on college students' self-esteem. *Journal of College Student Development*, *55*(5), 461–474. doi:10.1353/csd.2014.0051

National Science Foundation. (2017). *Women, Minorities, and Persons with Disabilities in Science and Engineering*. Author.

Norman, O., Ault, C. R., Bentz, B., & Meskimen, L. (2001). The black-white? achievement gapas a perennial challenge of urban science education: A sociocultural and historical overview with implications for research and practice. *Journal of Research in Science Teaching*, *38*(10), 1101–1114. doi:10.1002/tea.10004

Ong, M. (2005). Body Projects of Young Women of Color in Physics: Intersections of Gender, Race, and Science. *Social Problems*, *52*(4), 593–617. doi:10.1525p.2005.52.4.593

Ong, M., Wright, C., Espinosa, L. L., & Orfield, G. (2011). Inside the Double Bind: A Synthesis of Empirical Research on Undergraduate and Graduate Women of Color in Science, Technology, Engineering, and Mathematics. *Harvard Educational Review, 81*(Summer), 172–208. doi:10.17763/haer.81.2.t022245n7x4752v2

Owen, J., Tao, K., & Rodolfa, E. (2010). Microaggressions and women in short-term psychotherapy: Initial evidence. *The Counseling Psychologist, 38*(7), 923–946. doi:10.1177/0011000010376093

Pierce, C., Carew, J., Pierce-Gonzalez, D., & Willis, D. (1978). An experiment in racism: TV commercials. In C. Pierce (Ed.), *Television and education* (pp. 62–88). Beverly Hills, CA: Sage.

Seymour, E., & Hewitt, N. (1994). *Talking about leaving: Factors contributing to high attrition rates among science, mathematics & engineering undergraduate majors*. Final Report to the Alfred P. Sloan Foundation. Boulder, CO: University of Colorado, Ethnography and Assessment Research Bureau of Sociological Research.

Solorzano, D. G. (1998). Critical race theory, race and gender microaggressions, and the experience of Chicana and Chicano scholars. *International Journal of Qualitative Studies in Education: QSE, 11*(1), 121–136. doi:10.1080/095183998236926

Steele, C. M. (1997). A threat in the air: How stereotypes shape intellectual identity and performance. *The American Psychologist, 52*(6), 613–629. doi:10.1037/0003-066X.52.6.613 PMID:9174398

Steele, C. M., & Aronson, J. (1995). Stereotype threat and the intellectual test performance of African Americans. *Journal of Personality and Social Psychology, 69*(5), 797–811. doi:10.1037/0022-3514.69.5.797 PMID:7473032

Sue, D. W. (2010). *Microaggressions in everyday life: Race, gender, and sexual orientation*. Hoboken, NJ: Wiley.

Sue, D. W., Bucceri, J., Kin, A. I., Nadal, K. L., & Torino, G. C. (2007). Racial microaggressions and the Asian American experience. *Cultural Diversity & Ethnic Minority Psychology, 13*(1), 72–81. doi:10.1037/1099-9809.13.1.72 PMID:17227179

Sue, D. W., Lin, A. I., Torino, G. C., Capodilupo, C. M., & Rivera, D. P. (2009). Racial microaggressions and difficult dialogues on race in the classroom. *Cultural Diversity & Ethnic Minority Psychology, 15*(2), 183–190. doi:10.1037/a0014191 PMID:19364205

Torres, L., Driscoll, M. W., & Burrow, A. L. (2010). Racial microaggressions and psychological functioning among highly achieving African-Americans: A mixed methods approach. *Journal of Social and Clinical Psychology*, *29*(10), 1074–1099. doi:10.1521/jscp.2010.29.10.1074

Torres, L., Driscoll, M. W., & Burrow, A. L. (2010). Microaggressions and Psychological Functioning among High Achieving African-Americans: A Mixed-Methods Approach. *Journal of Social and Clinical Psychology*, *29*(10), 1074–1099. doi:10.1521/jscp.2010.29.10.1074

Wang, J., Leu, J., & Shoda, Y. (2011). When the seemingly innocuous "stings": Racial microaggressions and their emotional consequences. *Personality and Social Psychology Bulletin*, *37*(12), 1666–1678. doi:10.1177/0146167211416130 PMID:21885859

KEY TERMS AND DEFINITIONS

Microinsults: Communications demean a person's identity.

Microinvalidations: Nullification or negation of thoughts, feelings, and realities of specific groups of people.

Chapter 7 Secondary Education Mathematics and UDC Expectations:
Do They Align? If Not, What's Next?

Jeffery Fleming
University of the District of Columbia, USA

Shurron Farmer
University of the District of Columbia, USA

ABSTRACT

In this chapter, the authors wish to examine the alignment between Common Core mathematics standards and introductory mathematics courses at the University of the District of Columbia. In this study, the authors are not trying to determine the reasons students choose or not choose STEM-related fields as their majors; instead the authors are exploring the transition from secondary to post-secondary mathematics education by aligning the Common Core mathematics standards that have been the District of Columbia Public School (DCPS) system. The authors have observed from their teaching experiences that for many students, the transition from secondary to post-secondary mathematics has not been seamless. One factor that may cause a breakdown in this transition could be the misalignment between Common Core state standards mathematics content and the content of the introductory mathematics in the Division of Sciences and Mathematics at the University of the District of Columbia.

DOI: 10.4018/978-1-5225-7814-7.ch007

INTRODUCTION

In today's current secondary educational environment, the need for students to be exposed to science, technology, and engineering mathematics (STEM) runs high as science and technology continues to shape our society (National Research Council, 2011). Many young people, including those persons who are educated in urban environments across the nation, have strong aspirations of careers in the sciences, technology, mathematics, and engineering. However, their aspirations meet reality when they enter higher education and are exposed to the rigors of STEM. The authors' overall theme in this chapter surrounds the topic of whether secondary and post-secondary mathematics education need to align in order to ensure college readiness and, thus, support retention and success in mathematics degrees and fields.

The University of the District of Columbia (UDC) is, at once, very old and very new. The seeds of higher education for the District were planted in 1851 when Myrtilla Miner founded a "school for colored girls" in Washington, DC. Through a series of mergers among the District's teachers and technical colleges, a comprehensive university structure was envisioned for the city. On August 1, 1977, a public announcement was made of the consolidation of the District of Columbia Teachers College, the Federal City College, and the Washington Technical Institute into the University of the District of Columbia under a single management system. On the same day, the Board appointed Lisle Carleton Carter, Jr., the first president of the university. In 1999, the U.S. Department of Education formally designated UDC for recognition among the nation's Historically Black Colleges and Universities.

UDC is a Congressionally-mandated land-grant institution of higher education. It is a comprehensive public institution offering quality, affordable, postsecondary education at the certificate, associate, baccalaureate, and graduate levels. These programs prepare students for immediate entry into the workforce, for the next level of education, for specialized employment opportunities, and for lifelong learning.

UDC was built on the dreams of its founders, and it continues to transform itself to meet the changing needs of its students, enhance the lives of the community, and to meet the demands of the ever-changing landscape of the Nation's Capital: Washington, DC (Committee, 2016). Education is the key to equity--a fair shot at success. If you are born poor in the United States, you are most certain to die poor. Education offers just about the only escape there is from poverty. Employment training can help many begin climbing the income ladder by qualifying for better jobs. It can be a start, but the odds against entering the middle class are nearly insurmountable for workers who do not hold a bachelor's degree (Columbia, 2019-2022).

Higher education enrollment trends in the United States reflect an expanding college-age population. According to The National Center for Education Statistics (NCES), "Between 2000 and 2016, total undergraduate enrollment in degree-granting postsecondary institutions increased by 28 percent (from 13.2 million to 16.9 million students). By 2027, total undergraduate enrollment is projected to increase to 17.4 million students" (Hussar & Bailey, 2016). This increased enrollment is projected to come mainly from minority groups (i.e., Blacks and Hispanics) that have not traditionally studied STEM fields. However, despite enrollment growth and work-force demands, the Higher Education Research Institute (HERI) reports in their annual Survey of the American Freshman (2010) that only 1% to 3% of students intend to major in physical sciences, computer sciences, or mathematics/statistics, and that STEM majors consistently display a high switching rate of 40 to 60%. For mathematics majors, the switching rate tends to be very close to the 60% level. According to the Mathematics Association of America (2013), many students who originally intended to pursue STEM fields leave after their experiences in their introductory related courses.

In this chapter, the authors will examine the alignment between Common Core Stat Standards mathematics standards and introductory mathematics courses at the University of the District of Columbia. In this study, the authors are not trying to determine the reasons students choose or do not choose STEM-related fields as their majors; instead the authors are exploring the transition from secondary to post-secondary mathematics education by aligning the Common Core mathematics standards that have been adopted by the District of Columbia Public School (DCPS) system. The authors have observed from their teaching experiences that for many students, the transition from secondary to post-secondary mathematics has not been seamless. One factor that may cause a breakdown in this transition could be the misalignment between Common Core State Standards mathematics content and the content of the introductory mathematics in the Division of Sciences and Mathematics at the University of the District of Columbia. Thus, this initiative seeks to better understand the current landscape of STEM education at the K-12 level by exploring the alignment between K-12 mathematics requirements and course learning objectives in one public HBCU mathematics program.

To begin, the Common Core State Standards, as adopted by the District of Columbia, are reviewed. The Common Core State Standards Initiative states the following: "Building on the best of existing state standards, the Common Core State Standards provide clear and consistent learning goals to help prepare students for college, career, and life. The standards clearly demonstrate what students are expected to learn at each grade level, so that every parent and teacher can understand and support their learning. The standards are:

1. Research and evidence based
2. Clear, understandable, and consistent
3. Aligned with college and career expectations
4. Based on rigorous content and the application of knowledge through higher-order thinking skills
5. Built upon the strengths and lessons of current state standards
6. Informed by other top-performing countries to prepare all students for success in our global economy and society" (About the Standards, n.d.).

As of 2013, forty-one states, the District of Columbia, four territories, and the Department of Defense Education Activity (DoDEA) had adopted the Common Core State Standards (Standards in Your State, 2013). No set of grade-specific standards can fully reflect the vast variety of abilities, needs, learning rates, and achievement levels of students in any given classroom. Importantly, the standards provide clear guideposts along the way to support the goal of college and career readiness for all students.

According to the best available evidence, the mastery of each standard is essential for success in college, career, and life in today's global economy. While the standards set grade-specific goals, they do not define how the standards should be taught or which materials should be used to support students. States and districts recognize that there will need to be a range of supports in place to ensure that all students, including those with special needs and English language learners, can master the standards. It is up to the states to define the full range of supports appropriate for these students.

"For more than a decade, research studies of mathematics education in high-performing countries have concluded that mathematics education in the United States must become substantially more focused and coherent in order to improve mathematics achievement in this country," argues the *Introduction to the Common Core Math Standards*. "To deliver on this promise, the mathematics standards are designed to address the problem of a curriculum that is 'a mile wide and an inch deep'" (About the Standards, n.d.).

The Common Core State Standards (CCSS) is an effort to reform the education standards across the United States' educational system (Porter, McMaken, Hwang, & Yang). This initiative intended design is to create college and career ready students at the completion of high school. According to Tienken, some states jumped at the opportunity to participate in this movement without enough data to substantiate its claim of creating career and college ready students (Tienken, 2011).

Educational reform is not new in the United States educational system. Systematic changes and reform have been a topic of discussion since the 1980s (Delandshere & Petrosky, 2004). With the ever changing economy, locally and globally, students leaving secondary schools, forthcoming skillsets are needed not only to thrive, but to survive in this new economy which concerned not only the educators, but many business leaders, politicians and other key stakeholders as well (Mathis, 2010).

According to Mathis (Mathis, 2010), the Achievement Corporation, a company founded by the National Governors Association Center for Best Practices (NGA), was commissioned by NGA and the Council of Chief State School Officers to draft the new Common Core State Standards. Mathis (2010) goes on to suggest that Achieve Corporations created workgroups to create the new standards and these workgroups were comprised of individuals that were not all experts in the educational realm. As a result, a new set of common standards was established.

Many scholars have investigated the implication that the Common Core State Standards promote cognitive development and learning – that is, the function based on how a person processes and reasons information. This level of engagement in learning includes many components such as memory retention, problem-solving skills, thinking skills, and the perception of learned material that happens consciously and unconsciously (reference.com). Loveless (2013) examined the claims of cognitive development from the Common Core State Standards. In this engagement, Loveless arrived at the warning that the promises of the Common Core State Standards are ambiguous and too broad which may be a strategy used to garner support from the public. Thus, challenges exist in understanding the efficacy and appropriateness of the Common Core State Standards. Regardless, the standards have been adopted among many school districts and require intentional alignment with curricula.

Because of the Common Core State Standards Initiative, Common Core State Standards for Mathematics (CCSSM) began its work in July 2009 which based its content on "benchmarking of U.S. standards to high-performing countries on international exams, research on students' mathematical knowledge and skill from the National Mathematics Advisory panel, and research from students on student learning (Confrey, 2010).The authors of the CCSSM chose to make clear distinctions between content and process standards and did this by proposing the following eight mathematical practices. It is important to note that these eight mathematical practices are a combination of the National Council of Teachers of Mathematics' (NCTM) process standards and mathematical proficiencies as developed by National Research Council's (2001) report *Adding It Up*.

1. Make sense of problems and persevere in solving them
2. Reason abstractly and quantitatively
3. Construct viable arguments and critique the reasoning of others
4. Model with mathematics
5. Use appropriate tools strategically
6. Attend to precision
7. Look for and make use of structure
8. Look for and express regularity in repeated reasoning.

Table 1 lays out the mathematical content in grade K-5 in the CCSSM. Based on the changes put forth, increased emphasis is placed on fractions as numbers and on using the number line to build mathematical structure and visualization of the number system. Also, note that these standards deemphasized or removed algebra content from the early grades. The authors of CCSSM argued "(a) fewer changes means some must be eliminated, and (b) emphasizing number mastery would lead to quicker and more secure learning of these eliminated topics in middle grades" (Confrey, 2012). It must be said that many educators and scholars challenge this notion and will challenge this theory based on empirical data.

Table 2 details the mathematical content grade 6-8, the middle grades. Based on the content in elementary grades, grade 6 introduces ratio, proportion, percent, and statistics. Grade 7 presents the mathematical concept of rational numbers. One-third of the mathematical content of grade 8 consists of algebra topics for all students. It is noted by Confrey (2012) that one of the writers of the CCSSM, Phil Daro, proclaims that the CCSSM was written with the assumption of 100% mastery in all the preceding year's standards.

Table 3 describes the standards that exists for grades 9-12 via conceptual categories that schools must address. As outlined by Confrey (2012), major changes at the high school level are the masteries students must demonstrate beyond the traditional Algebra 2 content which includes such topics as periodic functions, polynomials, radicals,

Table 1. The CCSSM grades k-5 content

Content	Grades
Counting and Cardinality	K only
Operations and Algebraic Thinking	K-5
Number and Operations in Base Ten	K-5
Number and Operations – Fractions	3-5
Measurement and Data	K-5
Geometry	K-5

Table 2. The CCSSM grades 6-8 content

Content	Grades
Ratio and Proportional Relationship	6-7
The Number System	6-8
Expressions and Equations	6-8
Functions	8
Geometry	6-8
Statistics and Probability	6-8

Table 3. The CCSSM grades 9-12 content and conceptual categories

Conceptual Categories	
Number and Quantity	The Real Number System, Quantities, The Complex Number System, Vector and Matrix Quantities
Algebra	Seeing Structure in Expressions, Arithmetic with Polynomials and Rational Expressions, Creating Equations, Reasoning with Equations and Inequalities
Functions	Interpreting Functions, Building Functions, Linear, Quadratic, and Exponential Models, Trigonometric Functions
Modeling	
Geometry	Congruence, Similarity Right Triangles, and Trigonometry, Circles, Expressing Geometric Properties with Equations, Geometric Measurement and Dimension, Modeling with Geometry
Statistics and Probability	Interpreting Categorical and Quantitative Data, Making Inferences and Justifying Conclusions, Conditional Probability and the Rules of Probability, Using Probability to Make Decisions

advanced probability and statistics, and mathematical modeling. It is important to note that the writers of CCSSM strongly encourage that mathematical modeling be woven throughout all the other mathematical standards.

It should also be noted that the adoption of the Common Core State Standards means that states agree to implement, word-for-word, 100 percent of the English language arts and mathematical standards. This method of agreement ensures coherence across states to aid with students' mobility and provides rigorous content and application that is competitive in global society (CCSSI, 2010). Based on the argument that every student must master mathematical topics beyond Algebra 2 and that the CCSSM is a coherent model that draws on international models for mathematical practice, CCSSM is a strong push to improve mathematics achievement in this country (CCSSI, 2010) for which students are college and career ready.

Thus, the question remains: Are students enrolled in K-12 systems who have adopted and adhere to CCSSM college and career ready?

According the College and Career Readiness and Success Center, the District of Columbia defines college readiness as, " the level of preparation a student needs in order to enroll and succeed – without remediation – in a credit bearing course at postsecondary institution that offers a baccalaureate degree or transfer to a baccalaureate program, or in a high-quality program that enables students to enter a career pathway with potential future advancement" (College and Career Readiness and Success Center, 2018).

Many teachers, and of course some professors, have struggled with the age-old question in mathematics classrooms across the nation of "Why do I have to l learn this?" Being prepared for the next math course is not a sufficient answer to this question. Policymakers have begun to stress the need and usefulness for college and career readiness as the purpose of high school mathematics. High schools across the country are increasing the number of mathematics courses required for completion due to career and college readiness (Dossey, McCrone, & Halvorsen, 2016).

Catalyzing Change in High School Mathematics (Graham, 2018) gives the purposes of school mathematics:

The purposes of high school mathematics are not fundamentally different from the purposes of school mathematics across K-12 or education in general, but they need to be considered specifically with respect to high school mathematics education. These purposes have to do with empowerment. High school mathematics empowers student to -

- *expand professional opportunity;*
- *understand and critique the world; and*
- *experience wonder, joy, and beauty.*

Graham notes the different experiences students encounter when learning mathematics in kindergarten through grade 8 (K-8) versus their mathematical experiences in grades nine through twelve (9-12). In K-8, mathematics is clear in their everyday encounters in the world. From numeracy to geometry, K-8 mathematics exists in counting, sharing, quantifying, using money, sketching, drawing, and building. In grades 9-12, the language, concepts, and theory of mathematics is less evident in the daily lives of students. Thus, the multiple reasons for studying mathematics in high school are very important to pre-service teachers, current teachers and students (Graham, 2018).

Secondary Education Mathematics and UDC Expectations

For every student to be successful in learning mathematics at the high school level, critical conversations must be had between secondary and post-secondary mathematics educators and administrators. Five actions are recommended for post-secondary educators in *Catalyzing Change in High School Mathematics* (Graham, 2018):

1. Ensuring strong articulation and seamless pathways between the high school and the postsecondary mathematics curricula
2. Collaborating with school- and district-based mathematics educators
3. Working with in-service and preservice teachers to support research-informed and equitable instructional practices focused on the Essential Concepts
4. Collaborating with school and district educators to develop additional mathematics pathways and populating courses with Essential Concepts.
5. Collaborating with school and district educators to challenge and dismantle system structures that impede students' access to and success in mathematics (NCTM 2018).

The authors believe that post-secondary mathematics educators should use their expertise in the field to initiate recommendations such as the ones made by Graham. In collaborations with other experts in the field, at UDC the authors have initiated grants, summer workshops, and have worked with UDC mathematics faculty regarding its courses that service future K-12 mathematics educators to catalyze Graham's recommendations at UDC. *Catalyzing Change in High School Mathematics* (Graham, 2018) identifies the essential concepts stated in the above actions from the content areas of numbers, algebra and functions, statistics and probability, and geometry and measurement.

College readiness and completion justifies the need for alignment between secondary education and post-secondary educational student learning outcomes. UDC has been and continues to be a strong advocate for secondary education in the city of Washington, DC, especially since UDC is the nation's only public urban land-grant institution for higher education. UDC is aware that education is a major vehicle within the Black community regarding transition to the middle class. This raises a standard of living for the individual, the individual's family, and the community. This is a direct connection to UDC's new academic strategic plan – The Equity Imperative (2018-2023). Because of UDC's advocacy for secondary education in the city of Washington, DC, the authors share a passion for creating, in collaboration with others, a strong mathematics learning environment at UDC. The authors also believe such a strong learning environment at UDC will produce mathematicians who transition this passion to K-12 education environments in urban areas as well. Determined to provide high quality numeracy to our students, the mathematics program seeks to build knowledge and skills in the area of quantitative reasoning for STEM and non-STEM students.

The multivariate and multi-layered mission of the mathematics program at UDC – Van Ness Campus is threefold:

1. To create *producers* of mathematics; that is, to prepare degree-seeking students in mathematics with the knowledge, skills, and abilities to be creative thinkers and life-long learners as they pursue a career in mathematics, a mathematics-related field, or graduate studies.
2. To create *users* of mathematics; that is, to provide instructional support to all UDC departments requiring mathematics content for their majors.
3. To create *consumers* of mathematics; that is, to provide a variety of mathematics courses and sequences that satisfies the University's general education requirement in mathematics for every student.

The UDC mathematics program seeks to increase our students' numeracy and mathematical literacy, i.e. enhance one's ability to learn, refine, and exhibit mastery of mathematical concepts, skills and techniques, and apply mathematical techniques to solve real-life problems, and expand the scope of mathematics knowledge through research and collaboration with other scholars, universities, school districts and key stakeholders locally and beyond. To this end, the program has created the following student learning outcomes for mathematics:

1. Students will explore, state, define, and know mathematical language in the discipline area.
2. Students will apply mathematical algorithms, procedures in discipline areas.
3. Students will apply mathematical theorems, techniques, and strategies to problems in various disciplines.
4. Students will present oral and written mathematical concepts, procedures, results, and meaning of mathematical scenarios.
5. Students will use technology, appropriate software and vendor products to solve, analyze, interpret, and determine reasonableness of solutions in mathematical context and application.

As courses in Quantitative Reasoning, Pre-calculus and Calculus I often serve as "gateway courses" for UDC majors, the content in these courses is compared to CCSSM at the secondary level. The authors wish to:

1. Examine the mathematics curriculum and standards at the high school level for alignment with content in identified UDC mathematics courses.
2. Identify the major academic barriers impeding students' successful completion of courses in Quantitative Reasoning, Algebra, Geometry, or Pre-Calculus.
3. Examine the depth and breadth of the mathematics content in the CCSS as compared to UDC mathematics courses at the introductory level.

In the authors' teaching experiences at UDC over the past 10 years, they've seen how students' career aspirations potentially change based on the expectations of mathematics courses. Mathematics professors, and college professors in general, are called to set their standards, pedagogy, and practices so that students will achieve learning outcomes, be able to use course content meaningfully, and be exposed to the application of mathematics to situations not only in their careers as dictated by one's majors, but in everyday living as well.

According to Griffin, "National and private organizations have invested millions of dollars in science diversity programs designed to shift patterns of representation in science, technology, engineering, and math—the so-called STEM fields—in and beyond college. However, minoritized populations continue to be underrepresented in STEM disciplines" (Griffin, 2018). Many government agencies, corporations and businesses, continue to express their desire to diversify their STEM workforces through various funding initiatives and investments in secondary educational and higher education. At UDC, we strive to increase the number of persons completing degrees in STEM areas and increase interest of students to achieve advanced degrees in STEM. To insure appropriate approaches to achieving these goals, this chapter proposes to compare mathematics in the secondary educational system with expectations for mathematics at UDC. Understanding our place in history, embracing our presence and mission, under the guidance of the University's strategic initiatives, the authors believe in the necessity to minimize the challenges encumbered by first time college students during their transition from secondary to post-secondary educational environments.

METHODOLOGY

The authors examined the 2011 Common Core Math Crosswalk from the Washington, DC Office of the State Superintendent of Education (Education, 2011). In this Crosswalk, the authors noted the DC strands for mathematics student learning outcomes had already been aligned with CCSSM. Table 4 is a sample from the crosswalk demonstrating this alignment:

Table 4. DC math standard and Common Core crosswalk sample

Math		Grades Nine Through Twelve				
Priority Topic	**DC Strand**	**DC Standard**	**Grade Diff.**	**CC Matched Grade**	**CC Strand**	**CC Matched Standard**
	Algebra I: Number Sense & Operations Indicators	**9-12.N.AI.1** *(AI.N.1.) Use the properties of operations on real numbers, including the associative, commutative, identity, and distributive properties, and use them to simplify calculations.*		9-12	**Number & Quantity**	**9-12.N.RN.3** Use properties of rational and irrational numbers. Explain why the sum or product of rational numbers is rational; that the sum of a rational number and an irrational number is irrational; and that the product of a nonzero rational number and an irrational number is irrational.
				9-12	**Algebra**	**9-12.A.SSE.1** Interpret the structure of expressions. Interpret expressions that represent a quantity in terms of its context.*

The Crosswalk contains the DC standards for mathematics, the grade levels to which each of the standards applies, and the CCSSM. The alignment between the DC Standards and the CCSSM was based on the Revised Bloom's Taxonomy (Anderson, 2013) and course content. Bloom's Taxonomy is defined as an ordering of cognitive skills. This taxonomy had permeated teaching and instructional planning for almost 50 years before it was revised in 2001. The major differences lie in the more useful and comprehensive additions of how the taxonomy intersects and acts upon different types and levels of knowledge — factual, conceptual, procedural and metacognitive (Wilson, n.d.). In mathematics, metacognition is problem solving that uses the factual, conceptual, and procedural knowledge.

In a similar fashion, the authors began to align CCSSM to entry-level mathematics courses at UDC, particularly Quantitative Reasoning, Intermediate Algebra, Pre-Calculus, Calculus I, Discrete Mathematics, and Elementary Statistics I. The authors chose these courses because these courses are freshmen-level credit-bearing college courses that satisfy general education requirements at the University and simultaneously are gateway courses to STEM courses and degree programs at the University. The authors note that while at the Community College of the University of the District of Columbia (UDC-CC) there are mathematics courses (Basic Mathematics and Introductory Algebra) that prepare students for Intermediate Algebra, it is the hope and expectation at UDC that students enter mathematics courses without the need for mathematical remediation. Therefore, CCSSM is expected to demonstrate mastery of the course content of Basic Mathematics and Introductory Algebra.

ANALYSIS

The authors' alignment analysis of CCSSM with UDC mathematics courses is a multi-step process. The analysis includes the following:

1. Reviewing the CCSSM
2. Identifying the UDC mathematics courses to be included in the analysis
3. Identifying the student learning outcomes (SLOs) in the UDC mathematics courses that align with the CCSSM

For each CCSSM, the authors noted in which of the following courses the standard is either introduced, reinforced, or assessed for mastery:

- **Quantitative Reasoning (Quan. Reas.):** Quantitative Reasoning is an Interdisciplinary General Education course sequence that focuses on improving students' quantitative reasoning skills while exploring a given academic theme. The goal of this course is to teach students how to reason using the language and strategies of mathematics. Students analyze data, find connections among and between quantitative relationships, and communicate their findings using a variety of formats within different settings and to diverse audiences. By using a variety of strategies, students solve problems and analyze data in various contexts including applications demonstrating personal economic literacy.
- **Intermediate Algebra (Inte. Alg.):** Intermediate Algebra provides students with extension and refinement of concepts in algebra. Intermediate Algebra prepares students to be successful in Pre-Calculus, Finite Mathematics, or in Elementary Statistics I by extending the basic concepts introduced in Introductory Algebra. Simultaneously, this course provides time for the

students to properly absorb and apply the concepts, skills and attitudes required in higher level mathematics and math-related courses.

- **Pre-Calculus (Pre-Cal):** This course sequence is designed for students who are interested in disciplines such as computer science, engineering, mathematics or any disciplines requiring a firm understanding of algebraic and trigonometric concepts. The course is also designed for students who have successfully completed 2 years of high school algebra or equivalent college-level algebra courses and are desirous of enhancing their knowledge and skills of algebraic, logarithmic, exponential, and trigonometric functions and their graphs, and analytic geometry.
- **Finite Mathematics (Finite Math):** This course deals with pre-calculus topics in the general category of finite mathematics, beginning with a review of basic algebra and linear equations, and proceeding through logarithmic and exponential functions to prepare the student for applied calculus. The purpose of this course is to provide a strong foundation in college mathematics for business, economics, and social and life sciences that will serve in the study of business, economics, the social sciences, and the life sciences, all of which are heavily dependent and/or becoming more dependent upon mathematical theory and modeling.
- **Calculus I (Cal. I):** Calculus I can be described as an introduction to analysis. Knowledge of Calculus and its applications have been and are fundamental to the training of scientists, engineers, and mathematicians since its inception. This course sequence presents the essential topics of differential and integral calculus and other closely related topics in mathematics.
- **Discrete Mathematics (Dis. Math):** The course explores counting principles, logic, set theory, functions, properties of the natural numbers and of the integers, mathematical induction, complexity of algorithms, matrices, relations, and graph theory. The course is designed primarily for engineering and computer science students. A primary reason for studying the material in Discrete Mathematics is the abundance of applications one finds for it in the study of engineering and of computer science, such as modeling, graph theory, and the analysis of algorithms.
- **Elementary Statistics I (Elem. Stat. I):** This course introduces students to concepts and techniques of probability and statistics including: measures of central tendency and dispersion, probability and probability distributions, correlation and regression, statistical inference, and computer applications.

The authors then identified the student learning outcomes for the mathematics courses at UDC that aligned with the CCSSM. Table 5 is a sample of this analysis:

Table 5. UDC mathematics program student learning outcomes

DISCIPLINE AREA	UDC Mathematics Program SLO (Student Learning Outcome)
Intermediate Algebra	Solve algebraic equations and inequalities
Intermediate Algebra	Solve applied problems involving algebraic equations, inequalities, and functions
Intermediate Algebra	Solve applied problems involving exponential and logarithmic equations, inequalities, and functions
Intermediate Algebra	Solve exponential and logarithmic equations and inequalities
Functions	Modeling using functions
Functions	Define relation, function, and inverse
Functions	Describe a function as either one-to-one, onto, or invertible
Functions	Perform arithmetic operations with functions
Functions	Sketch and analyze various representations of functions
Statistics	Calculate the relationship between explanatory variables and response variables
Statistics	Know how to gather, organize and display data via observation and experiment
Statistics	State the probability distributions of discrete and continuous random variables
Statistics	State types of statistical inference

Next, the authors compared the CCSSM to the UDC student learning outcomes for alignment and observed which CCSSM aligned with student learning outcomes in UDC mathematics courses. The authors noted the UDC courses where the standards are either introduced, refined, or assessed for mastery. Table 6 is a sample of this comparison.

The authors note that although there are many UDC student learning outcomes that don't align with the CCSSM, such observation should not be perceived in a negative manner. It's the opinion of the authors that the CCSSM are designed to prepare students in such a way that students should enter college mathematics courses prepared to succeed in such courses without the need for remediation at the college level.

Finally, the authors identified the UDC courses where the UDC matched student learning outcomes that aligned with CCSSM are explored. The chart in Table 7 displays a sample of this alignment.

Table 6. Alignment of Common Core matched standards with UDC matched student learning outcomes (SLO) - Sample

Common Core Matched Standard	UDC Matched SLO
9-12.A.CED.1 Create equations that describe numbers or relationship. Create equations and inequalities in one variable and use them to solve problems. Include equations arising from linear and quadratic functions, and simple rational and exponential functions.	Solve algebraic equations and inequalities
9-12.F.LE.5 Construct and compare linear, quadratic, and exponential models and solve problems. Interpret the parameters in a linear or exponential function in terms of a context.	No alignment with UDC Matched SLO
9-12.N.RN.1 Extend the properties of exponents to rational exponents. Explain how the definition of the meaning of rational exponents follows from extending the properties of integer exponents to those values, allowing for a notation for radicals in terms of rational exponents. For example, we define 5^(1/3) to be the cube root of 5 because we want [5^(1/3)]^3 = 5^[(1/3) x 3] to hold, so [5^(1/3)]^3 must equal 5.	Modeling Using Functions
9-12.F.IF.1 Understand the concept of a function and use function notation. Understand that a function from one set (called the domain) to another set (called the range) assigns to each element of the domain exactly one element of the range. If f is a function and x is an element of its domain, then f(x) denotes the output of f corresponding to the input x. The graph of f is the graph of the equation y = f(x).	Define relation, function, and inverse
9-12.F.IF.4 Interpret functions that arise in applications in terms of the context. For a function that models a relationship between two quantities, interpret key features of graphs and tables in terms of the quantities, and sketch graphs showing key features given a verbal description of the relationship. Key features include: intercepts; intervals where the function is increasing, decreasing, positive, or negative; relative maximums and minimums; symmetries; end behavior; and periodicity.	Sketch and analyze various representations of functions

The Mathematics Program has identified 47 student learning outcomes for its mathematics courses. Fifteen of these outcomes aligned with the Common Core Crosswalk. The bar chart in Figure 1 displays the percentage of CCSSM that align with student learning outcomes in the UDC courses:

The authors make several observations based on this analysis. There are 104 mathematics standards in the Common Core Crosswalk. These standards were aligned to CCSSM up to 492 times. The CCSSM aligned with 15 UDC mathematics student learning outcomes. The majority of CCSSM align with Intermediate Algebra, Finite Mathematics, and Pre-Calculus. There were 40 occurrences of CCSSM that aligned with the UDC discipline area Intermediate Algebra. These standards aligned with 3 UDC student learning outcomes. The occurrences appeared in the following UDC math courses: Quantitative Reasoning, Intermediate Algebra, Pre-Calculus, Finite Math, Calculus I. As displayed in Table 8, the largest percentage alignment of the CCSSM is in the UDC math courses Intermediate Algebra, Finite Math, Pre-Calculus. The

Table 7. UDC matched SLOs and the courses where they are explored

UDC Matched SLO	Inte. Alg.	Fin. Math	Pre-Cal.	Cal. I	Dis. Math	Elem. Stat. I	Quan. Reas.
Solve algebraic equations and inequalities	X	X	X	X			X
Solve applied problems involving algebraic equations, inequalities, and functions	X	X	X				
Solve applied problems involving exponential and logarithmic equations, inequalities, and functions	X	X	X				
Solve exponential and logarithmic equations and inequalities	X	X	X				
Modeling using functions	X	X	X	X		X	X
Define relation, function, and inverse	X	X	X	X	X		
Describe a function as either one-to-one, onto, or invertible	X	X	X	X	X		
Perform arithmetic operations with functions	X	X	X	X			
Sketch and analyze various representations of functions	X	X	X	X			
Calculate the relationship between explanatory variables and response variables						X	
Know how to gather, organize and display data via observation and experiment						X	
State the probability distributions of discrete and continuous random variables						X	
State types of statistical inference						X	

smallest percentage alignment is in the UDC math courses Quantitative Reasoning, Calculus I, Elementary Statistics I, and Discrete Math. Finally, in examining the Common Core Crosswalk, the authors noted that DC Standards were broadly stated, while the CCSSM were more content driven.

For example, the following DC Mathematics Standard states:

Figure 1. Common Core UDC alignment statistics

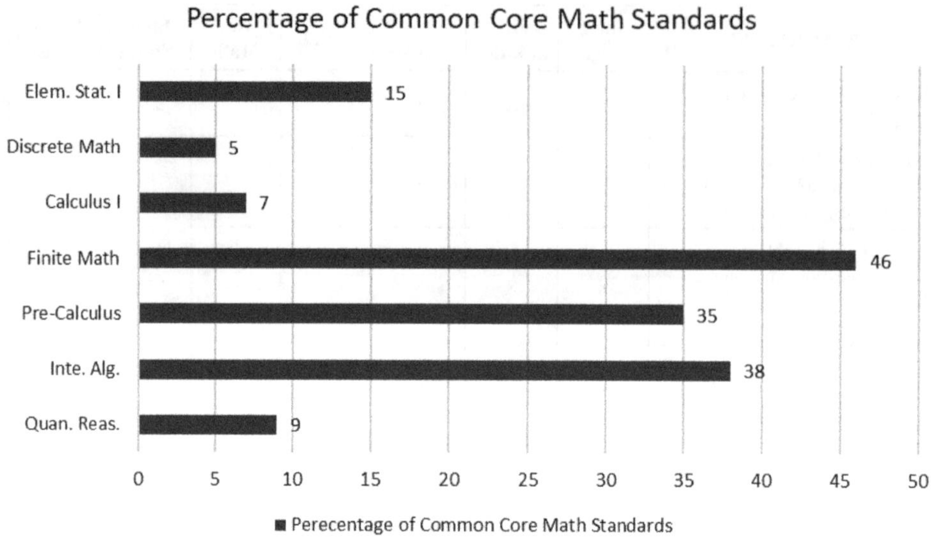

9-12.N.AI.4 (AI.N.4.) Use estimation to judge the reasonableness of results of computations and of solutions to problems involving real numbers, including approximate error in measurement and the approximate value of square roots. (Reminder: This is without the use of calculators.) (Initiative, 2010)

This standard is aligned to the Common Core State Standard:

9-12.F.LE.5 Construct and compare linear, quadratic, and exponential models and solve problems. Interpret the parameters in a linear or exponential function in terms of a context. (Initiative, 2010)

The Common Core Standard aligns with the UDC student learning outcome:

Modeling with Functions

The Common Core Standard is introduced, refined, or assessed for mastery in the UDC course Intermediate Algebra.

DISCUSSION

While CCSSM focus solely on mathematical content, the authors' analysis and Table 8 suggests more discussion needs to happen regarding CCSSM in an urban learning environment and its impact on mathematics education at UDC. The Office of the State Superintendent of Education (OSSE) states that it is

the State Education Agency for the District of Columbia charged with raising the quality of education for all DC residents. OSSE serves as the District's liaison to the U.S. Department of Education and works closely with District's traditional and public charter schools to achieve its key functions:

1. *Overseeing all federal education program and related grants administered in the District of Columbia.*
2. *Developing state-level standards aligned with school, college, and workforce readiness expectations.*
3. *Ensuring access to high-quality child care and universal pre-kindergarten for eligible District families.*
4. *Providing resources and support to assist the District's most vulnerable student populations.*
5. *Administering the annual Partnership for Assessment of Readiness for College and Careers (PARCC), the statewide student academic achievement exam.*
6. *Providing regional, door-to-door transportation to school for District children with special needs.*
7. *Awarding higher education financial assistance to eligible District students at public and private colleges and universities in DC and across the country.*
8. *Increasing health and physical education awareness as well as ensuring access to free meals year-round.*
9. *Overseeing the DC State Athletic Association (DCSAA), which provides interscholastic athletic programming that enriches the education experiences of all student-athletes.*
10. *Providing a one-stop source of statewide school data on each traditional and public charter school as well as resources to support children from birth to post-secondary education: LearnDC. (Office of the State Superintendent of Education, 2018)*

As part of the state education system for the District of Columbia, OSSE serves:

- *92,245 students attending all K-12 schools*
- *68 local education agencies, including DC Public Schools and charter organizations*
- *264 schools, both DCPS and charter*
- *52.7 percent of students attend DC Public Schools*
- *47.3 percent of students attend charter schools. (Office of the State Superintendent of Education, 2018)*

Additionally, the statistics for enrollment by race and ethnicity of students are as follows:

- 68.1 percent of students identify as African-American/Black (non-Hispanic)
- 18.1 percent of students identify as Hispanic/Latino
- 10 percent of students identify as White (non-Hispanic)
- 1.6 percent of students identify as Asian
- 2 percent of students identify as multiracial
- Less than one percent of students identify as Native American/Alaskan
- Less than one percent of students identify as Pacific Islander/Native Hawaiian (Office of the State Superintendent of Education, 2018).

In an educational system in an urban environment with more than 86 percent of students who identify as either African-American/Black (non-Hispanic) or Hispanic/Latino, data suggests a significant difference between increases in academic success on PARCC among these ethnic groups as compared to White/Caucasian students. The data in Table 8 illustrates that students in nearly all race/ethnicity groups saw improvement in both English Language Arts (ELA) and math from 2017 to 2018. Gaps in performance across these groups persist.

Thus, while CCSSM focus solely on mathematical content, the authors' analysis and Table 8 suggests more discussion needs to happen regarding CCSSM in an urban learning environment with UDC. Lou Matthews suggests the definition of "urban" in mathematics education should be defined in a context not confined to geography. Matthews writes, "… the view of urban domain extends beyond the geographical context, into the lives of people within the multitude of cultural, social, and political spaces in which mathematics teaching and learning takes place" (Matthews, 2008).

Table 8. Results for race/ethnicity groups improved

Race/Ethnicity	State ELA % 4+			State Math % 4+		
	2017	2018	% Point Change	2017	2018	% Point Change
All Groups	30.5%	33.3%	2.8%	26.9%	29.4%	2.5%
Asian	66.2%	71.6%	5.4%	64.5%	69.8%	5.3%
Black/African-American	22.0%	24.7%	2.7%	18.6%	20.7%	2.1%
Hispanic/Latino of any race	28.9%	32.0%	3.1%	26.0%	28.2%	2.2%
Two or More Races	66.7%	65.7%	-1.0%	62.8%	63.9%	1.1%
White/Caucasian	82.0%	82.1%	0.1%	75.5%	78.8%	3.3%

(Presentation, 2018)

The authors' analysis between CCSSM and UDC mathematics student learning outcomes demonstrated that the 492 occurrences of CCSSM seamlessly aligned with 15 UDC mathematics student learning outcomes. From this analysis, the authors wish to further examine this implementation and execution of CCSSM under Matthews' definition of "urban learning" (Matthews, 2008). A discussion around CCSSM and college readiness for UDC in mathematics could begin with urban learning of mathematics around the 15 UDC mathematics student learning outcomes that aligned with CCSSM.

LIMITATIONS

Although the UDC mathematics program continues to improve its engagement with students to be producers, consumers, and users of mathematics, the authors' alignment between CCSSM and UDC student learning outcomes was based on standards from Academic Year 2014-15. The authors noted that the crosswalk between DC standards and the newly refined CCSSM is problematic due to the repetitive nature of the CCSSM in the crosswalk. Lastly, CCSSM and UDC mathematics standards learning outcomes may be explored differently based on teachers' philosophy and pedagogy, which could limit students' conceptual and procedural knowledge of mathematical content.

CONCLUSION AND NEXT STEPS

From the analysis of the alignment of CCSSM with student learning outcomes of UDC mathematics courses, in the future the authors wish to: examine the impact of CCSSM on the UDC student population; compare students who enter UDC from schools that adopted CCSSM to students who entered UDC from schools that did not adopt CCSSM to see if there a significant difference between the two populations; for UDC mathematics courses that displayed at least 30% alignment with CCSSM, examine the PARCC scores and/or final grades from students' high school mathematics courses to determine if there's a correlation and predict student success; create and strengthen partnerships with UDC education faculty to improve the exploration of CCSSM, and teaching and training of pre-service and in-service mathematics teachers in K-12 urban education environments; and establish a partnership with school and district K-12 educators, particularly educators in K-12 urban learning environments, to challenge and dismantle system structures that impede students' access to and success in mathematics.

REFERENCES

About the Standards. (n.d.). Retrieved 5 29, 2018, from Common Core State Standards Initiative: http://www.corestandards.org/about-the-standards

Anderson, L. e. (2013). *A Taxonomy for Learning, Teaching, and Assessing: A Revision of Bloom's Taxonomy of Educational Objectives, Abridged Edition*. Pearson.

CCSS Math - Common Core State Standards Initiative. (n.d.). Retrieved 5 29, 2018, from Common Core State Standards Initiative: http://www.corestandards.org/math

College and Career Readiness and Success Center. (2018). Retrieved from https://ccrscenter.org/ccrs-landscape/state-profile/district-columbia

Committee, U. S.-S. (2016). *Connecting the Dots: Sustaining a Culture of Assessment*. Washington, DC: UDC Self-Study Report, Prepared for the Middle States Commission On Higher Education and the UDC community.

Confrey, J., & Krupa, E. E. (2010). *Curriculum Design, Development, and Implementation in an Era of Common Core State Standards: Summary Report of a Conferent*. Arlington, VA: Center for the Study of Mathematics Curriculum.

Confrey, J., & Krupa, E. E. (2012). The Common Core State Standards for Mathematics: How Did We Get Here, and What Needs to Happen Next? In C. R. Hirsch, G. T. Lappan, & B. J. Reys (Eds.), Curriculum Issue in an Era of Common Core State Standards for Mathematics (pp. 3-15). The National Council of Teachers of Mathematics.

Delandshere, G., & Petrosky, A. (2004). Political rationales and idelogical stances of the standards-based reform of teacher education in the U.S. *Teaching and Teacher Education*, *20*(1), 1–15. doi:10.1016/j.tate.2003.09.002

Dossey, J. A., McCrone, S. S., & Halvorsen, K. T. (2016). *Mathematics Education in the United States, 2016: A Capsule Summary Fact Book*. Reston, VA: National Council of Teachers of Mathematics.

EducationD. O. (2011, June 6). Retrieved from https://osse.dc.gov/publication/math-common-core-crosswalk

Graham, K. J. (2018). *Catalyzing Change in High School Mathematics: Initiating Critical Conversations*. Reston, VA: The National Council of Teachers of Mathematics.

Griffin, K. A. (2018, April 23). *Addressing STEM Culture and Climate to Increase Diversity in STEM Disciplines.* Retrieved from Higher Education Today: https://www.higheredtoday.org/2018/04/23/addressing-stem-culture-climate-increase-diversity-stem-disciplines/

Hussar, W. J., & Bailey, T. M. (2016). *Projection of Education Statistics to 2024 (NCES 2016-013).* Washington, DC: U.S. Department of Education, National Center for Education Statistics.

Initiative, C. C. (2010). *Common Core State StandardS.* Retrieved 5 29, 2018, from Common Core State Standards Initiative: http://www.corestandards.org/assets/CCSSI_Math%20Standards.pdf

Loveless, T. (2013). *How well are American students learning? With sections on the latest international tests, tracking, and ability grouping, and advanced math 8th grade.* The Brown Center on Educational Policy.

Mathis, W. (2010). *Education and the Public Interest Center & Education Policy Research Unit.* Retrieved from http://epicpolicy.org/publication/common-core-standards

Matthews, L. E. (2008, December). Illuminating Urban Excellence: A Movement of Change with Mathematics Education. *Journal of Urban Mathematics Education, 11*(1-2), 7–10. http://education.gsu.edu/JUME Retrieved December 14, 2018

Office of the State Superintendent of Education. (2018). *Office of the State Superintendent of Education.* Retrieved from Office of the State Superintendent of Education: https://osse.dc.gov/page/about-osse

Porter, A., McMaken, J., Hwang, J., & Yang, R. (n.d.). Common Core State Standards: The new U.S. intended curriculum. *Educational Researcher, 40*(3), 103-115.

Presentation, 2.-1. D. (2018). *Office of State Superintendent of Education.* Retrieved from https://osse.dc.gov/sites/default/files/dc/sites/osse/page_content/attachments/2018%20PARCC%20Results%20Release%20%28Aug.%2016%2C%20 2018%29.pdf

Standards in Your State. (2013). Retrieved from Common Core State Standards Initiative: http://www.corestandards.org/standards-in-your-state

Tienken, C. (2011). *CCCS: I wonder? Kappa Delta Pi Record.* Retrieved from ERIC database. (EJ939062)

University of the District of Columbia. (n.d.). *The Equity Imperative: A Strategic Plan to Regenerate the University of the District of Columbia As A Public Higher Education Model of Urban Student Success.* Washington, DC: University of the District of Columbia.

Wilson, L. (n.d.). *The Second Principle - The Work of Leslie Wilson, Ed. D.* Retrieved from The Second Principle: https://thesecondprinciple.com/teaching-essentials/beyond-bloom-cognitive-taxonomy-revised/

Chapter 8
Improving STEM Career Aspirations in Underrepresented Populations:
Strategies for Urban Elementary School Professionals

Amanda J. Rockinson-Szapkiw
University of Memphis, USA

Logan R. Caldwell
University of Memphis, USA

ABSTRACT

The United States preK-12 educational system is uniquely positioned to assist elementary students, especially girls and minorities, in experiencing achievement and developing a self-efficacy in STEM, and consequently producing students who pursue STEM degrees in college and STEM careers. Thus, this chapter uses Gottfredson's Theory of Circumscription and Compromise and Bandura's concept of self-efficacy to identify barriers to STEM career aspirations that girls and minorities face in urban settings. These theories and research are also used to identify strategies for urban teachers and professional school counselors to improve elementary girls and minorities' STEM career aspiration.

DOI: 10.4018/978-1-5225-7814-7.ch008

INTRODUCTION

Encouraging students to pursue degrees and careers in Science, Technology, Engineering, and Mathematics (STEM) is vital to the United States for both economic and ecological reasons (Douglas & Strobel, 2015). Having individuals in STEM fields influences the United States' ability to remain globally competitive in a world that is becoming increasingly dependent upon science and technology (Harwell et al., 2015; Museus, Palmer, Davis, & Maramba, 2011; National Academy of Science, 2006). Developing scientists and engineers for the 21st century is vital in meeting needs for clean water, adequate food supply, global communication, and shelter (Kanwar, 2010) as well as in enhancing the globalization of state-of-the-art-technologies (Buschor, Berweger, Frei, & Kappler, 2014).

Projections have demonstrated that nine out of ten of the fastest growing occupations require at least a bachelors' degree with mathematical or scientific training, and the largest increases in job openings in the United States are in computer science and engineering fields (United States Department of Labor, USDOL, 2013). Between 2008- 2018, STEM jobs increased by 17% in comparison to non-STEM careers increasing by only 9.8%. Additionally, individuals in STEM careers earn on average 26% more than those in non-STEM careers (Langdon, McKittrick, Beede, Khan, & Doms, 2011). Unfortunately, the Bureau of Labor Statistics (BLS) (2010) has noted that the United States is having an increasingly difficult time filling STEM positions due to retirement trends and a decrease in students choosing STEM degrees in college. In the 21st century, only 20% of the United States workforce has the STEM skills needed for new jobs (Granovskiy, 2018; NSTC, 2011).

Underrepresentation of women and minorities in STEM fields is also a concern (NSF, 2009, 2015). In the fastest growing fields of computer science and engineering, women hold fewer than twenty-five percent of the jobs (USDOL, 2013). Beginning at a young age, in general, men demonstrate a higher interest in the science-related study than women (Caleon & Subramanian, 2008). Despite similar capabilities, girls perform poorer than boys in science and math (Smith, Sansone, & White, 2007). Furthermore, African-Americans hold about only eight percent of science and engineering occupations; Hispanics only four percent; and American Indians, Native Alaskans, Native Hawaiians, Other Pacific Islanders, and multiracial individuals only three percent (NSF, 2009, 2015).

Complex factors influence the lack or skilled STEM workers as well as the underrepresentation of women and minorities in STEM careers. Among these factors are stereotypes and self-efficacy. Gender stereotypes that children encounter in early elementary school have been theorized to influence children's perceptions of self and gender appropriate careers in STEM (Gottfredson, 2002; Smith, Sansone, &

White, 2007). Similarly, racial and economic minority children in elementary school may develop negative stereotypes about their social groups' academic competence (McKown & Weinstein, 2003), which interacts with their developing beliefs about what type of jobs are appropriate for them (Gottfredson, 2004; Steele, 1997). Lack of self-confidence in STEM subjects and low-STEM career self-efficacy has also been attributed to minorities and girls lack STEM-related occupational aspirations (Grossman & Porche, 2013; Shoffner, Newsome, Barrio Minton, & Wachter Morris, 2015).

The United States urban preK-12 educational system is thus uniquely positioned to assist elementary students, especially girls and minorities living in poverty, in addressing negative stereotypes and improving self-efficacy related to STEM, thus, consequently producing students who pursue STEM degrees in college and STEM careers. While research on the effectiveness of STEM educational strategies and holistic programs in urban elementary school remains limited (Harwell et al., 2015; Scantlebury, 2014; Varelas, Kane, & Wylie, 2011), there are theories and research that are useful to inform strategies that urban school teachers and professional school counselors can implement to increase the STEM interest and pursuit of STEM degrees among girls and minorities. This chapter will use Gottfredson's (1981) Theory of Circumscription and Compromise to identify barriers to STEM career aspirations that girls and minorities face in urban settings. Strategies will be presented for urban teachers and school counselors to improve elementary girls and minorities' STEM career aspiration.

GOTTFREDSON'S (1981) THEORY OF CIRCUMSCRIPTION AND COMPROMISE

While there has been some critique about the theories comprehensiveness and empirical validation, Gottfredson's (1981) Theory of Circumscription and Compromise provides insight on the role of teachers and professional school counselors in urban settings play in helping elementary students to combat stereotypes that may inhibit them from STEM career aspirations and achievement. For, individuals pursue careers that are congruent with their self-concept and reflect their understanding of different occupations (Gottfredson, 1981). An individual's self-concept or self-creation is defined as the beliefs they hold about themselves and is determined by gender and racial role socialization, socioeconomic status, values, abilities, and interest (Gottfredson, 2004). Self-concept is created through a dynamic interaction between genetic makeup and the environment. Individuals choose careers in alignment with the self-concept they create.

Gottfredson (2004) explained career choice as a developmental process that begins in childhood, where children begin to synthesize and organize complex occupational information beginning at the age of three with this process becoming more complex as children progress in age and intelligence. She hypothesized that career decisions are made in four stages, with the elimination of inappropriate careers (called circumspection) occurring at each stage (Gottfredson, 2004).

In the first stage, occurring between ages three and five, children begin to orient toward size and power. They recognize that occupations are adult roles, and they begin eliminating fictional characters and animals from their career choices (Gottfredson, 2004). For example, Alexandria decides that being a bunny is no longer a viable career choice but being a mathematician is.

In the second stage (ages six to eight), children become aware of gender stereotypes associated with occupations. Men and women are recognized as holding different types of jobs, and children identify with either male or female occupations based on their sex. Roles are seen as distinct, and cross-gendered occupations are resented (Gottfredson, 2004). During this stage, many girls begin to perceive science and engineering careers as masculine as they are exposed to pictures and experiences demonstrating that scientists and engineers are older, white men (Mead & Metraux, 1957; Finson, Pedersen, & Thomas, 2006; Türkmen, 2008). Thus, by the age of eight, girls may eliminate many STEM careers from their vocational options based on their sex categorizations.

Societal valuation and prestige are characteristics of the third stage (Gottfredson, 2004). Children and adolescence, ages nine and thirteen, categorize jobs based on social status (e.g., income, education, lifestyle, race, living environments, culture) and sex types. Their vocational choices are narrowed based upon wanting to belong, be respected, and be comfortable. Based on social interactions, they begin to identify their place in the social order and eliminate career options based on a minimum status level and a maximum status level. This is especially significant in urban settings as approximately one-third of United States citizens live in urban areas, and children living in urban areas are more likely to come from low-income families than children in rural or suburban areas (U.S. Census Bureau, 2012). Many children from urban areas live in poverty. By the end of this stage, children and adolescence rule out numerous occupations for being the wrong sex type or coming from unacceptably low income or high-income families. For example, Chandler is an African American boy growing up in urban Chicago. He has a single mother who is working two jobs, but his family still lives in poverty. At the age of six, he wanted to be an engineer. Through socialization at school, he has ruled out engineering as a career. He believes it is too hard for a boy from his area to obtain a college degree, and he has learned that those from his community who go to college are rejected by key community members. For, individuals who go to college think of themselves as, "too good."

In the final stage (ages 14 and above), complex concepts such as interests, abilities values, work-life balance, and personality are taken into consideration and used to exclude options to identify an appropriate field of work. During this stage, adolescents and young adults are more conscious about their career search and evaluate their traits, their abilities, and willingness to struggle with specific societal barriers to achieve their career aspirations (Gottfredson, 2004). Depending upon determination and self-concept, career options are further eliminated. Due to the process of gradual elimination, which is often inadvertent in the first three stages, elementary students have eliminated many careers not congruent with whom they understand themselves to be. Moreover, it is during elementary school that gender-based, race and social status elimination occurs. By the end of elementary school, many students, especially girls and minority students living in poverty in urban areas, have ruled out many STEM careers as occupations that are too masculine or do not align with their social status (e.g., requires too much expensive schooling) or color. The stages of circumscription are processes by which students eliminate occupations deemed unacceptable based on criteria discussed; however, compromise, which is the process of relinquishing preferred career choices for ones that are less acceptable or compatible but are accessible, often further limits career choices. The short-list of career choices a student has based on perceived size, power, societal valuation, prestige and gender, is further narrowed by social space (i.e., what is available within a certain urban area where the student attends school or what is available to a student whose parents do not have money for him or her to attend college). The gender, race, and social status selections, compromises, and elimination of occupations through stages have been well supported by empirical research (Blanchard &Lichtenberg, 2003; Gottfredson & Lapan, 1997; Hannah & Kahn, 1989; Helwig, 1998).

STEM STEREOTYPES IN URBAN ELEMENTARY SCHOOL

In alignment with Gottfredson's (2004) theory, decades of research have shown that children develop stereotypical belief systems about the types of people that work in STEM fields. Mead and Metaux (1957) analyzed essays high students wrote about scientists and found that students held stereotypical views of scientists as Caucasian males who work in laboratories. Replication studies and similar variations of this study over decades continue to have the same result; high school, middle school, and elementary school students continue to perceive scientists and other STEM professionals as Caucasian men (Bodzin & Gehinger, 2001; Etzioni & Nunn, 1974; Mason, Kahle, & Gardner, 1991; Ozel, 2012). Examining 247 American children,

Cvencek, Meltzoff, and Greenwald (2011) found that, as early as second grade, children hold the stereotype that math is a male field. Moreover, McKown and Weinstein (2003) found elementary aged children, ages 6 to 10, held negative views about the academic competence of racial minorities, especially in STEM-related disciplines. Shenouda and Danovitch (2014) examined children ages 4 to 10 and found that gender stereotypes occur as early as 4 years old and affect female STEM performance and belief that STEM is for boys. These incorrect stereotypes negatively influence elementary students' STEM achievement (Huguet & Régner, 2007; McKown & Weinstein, 2003; McKown & Strambler, 2009), children's interest in science careers (Buldu, 2006), and, ultimately, interest and attitudes toward STEM fields by the time they get to college and pursue careers as adults (Forbes & Schmader, 2010; Morgan, Isaac, & Sansone, 2001). Occupational aspirations are strongly associated with images children develop at young ages, and incorrect, stereotypical perceptions of STEM professionals can inhibit girls and racial minorities from urban areas from pursuing careers in STEM (Osborne & Collins, 2001; Capobianco, Diefes-dux, Mena, & Weller, 2011; Grossman & Porche, 2013; Master & Meltzoff, 2016).

SELF-EFFICACY

Self-efficacy has also been a vital factor in predicting the STEM interest and career aspirations of girls and underrepresented minorities (Grossman & Porche, 2013). Self-efficacy is individuals' perceptions about their capabilities within a specific area and was introduced by Bandura (1977, 1997). Bandura (1997) theorized that individuals acquire their self-efficacy from four primary sources: (1) experiencing mastery (e.g. achieving a goal), (2) vicarious (modeled) experiences, (3) social or verbal persuasion by a significant person (e.g. a teacher communicating a belief that a student can be a successful engineer), and (4) physiological or emotional responses (e.g. anxiety or excitement about a math class). Students with comparable intellectual ability can perform differently based on their self-perceptions of their abilities. Students with strong self-efficacy in STEM perceive they can succeed in STEM courses and careers, and their self-efficacy can influence the STEM career choices they make.

Self-efficacy related to understanding the STEM beliefs and career choices, especially for girls and underrepresented minority students, is further understood in the Social Cognitive Career Theory (SCCT) developed by Lent et al. (1994), which is based on Bandura's Social Cognitive Theory. Lent et al. (1994) posited that the interaction influences career choices among outcome expectancies (e.g., beliefs about the consequences or outcomes of action), interests or goal, and self-efficacy. A student's choice to pursue or not pursue a STEM career is closely related to the three areas of self-efficacy (e.g., belief about the ability to be successful in STEM), outcome expectations (e.g., belief about consequences pursuing a STEM career), and personal interests. So, even when a student experienced a previous success with a STEM task or course, a perceived barrier (e.g. perceived lack financial means to attend a STEM club or pursue a STEM degree), which are often perceived as children learn about societal valuation and prestige (e.g. third stage of Gottfredson(2004)) can result in negative outcome expectations and lowered self-efficacy, which may ultimately influence the choice to avoid a STEM career (Shoffner, Newsome, Barrio, Minton, & Wachter Morris, 2015). In Gottfredson's (2014) second stage (ages six to eight), a student is succeptable to encounter stereotypes (e.g., girls can't succeed in math) such as those related to gender, ethnicity, and socioeconomic status which can also negatively influence self-efficacy in STEM, and ultimately, alter career choices. One's attitudes about their abilities are a significant factor in choosing a career and determining the likelihood of future acceptance and success in that career. Interventions, such as Program for Research Initiatives in Science and Math (PRISM), however, can be employed in urban settings to counteract perceived barriers and stereotypes, and thus, positively influence STEM career self-efficacy and assist students in visualizing themselves as successful STEM workers. Carpi, Ronan, Falconer, and Lents (2017, p. 189) explained,

By setting high expectations for students and expecting them to commit significant time over multiple years toward the pursuit of research, students become successful in scientific endeavors and begin to consider careers that they never did before or that they may have dismissed in their earlier thinking. Through interventions aligned with SCCT targeting career knowledge, self-efficacy, and overcoming barriers, the PRISM program illustrates that previously foreclosed or unimagined STEM career pathways can become targets and interests for students at Minority-Serving Institutions.

STRATEGIES

Therefore, to diversify the STEM workforce, it is essential as Gottfredson (1981, 2004) purported to provide systematic, diverse STEM exposure and experience for elementary students in urban areas to broaden their range of occupational choices and break down gender, race, and class stereotypes; and elementary school teachers and counselors can play a significant role in providing such exposure and experience. Moreover, as Lent et al. (1994) suggested, it is also important to provide experiences for students to develop their self-efficacy and develop positive outcome expectations. Drawing from the research, teachers and professional school counselors can use the following strategies to address self-identity and self-efficacy in STEM for underrepresented populations.

Strategy 1: Seek Training on Providing Equitable Access by Understanding Students' Cultures and Backgrounds

Understanding students' cultures and backgrounds is a necessity in providing equitable and accessible STEM opportunities for students. Gottfredson (1981) stated that all individuals pursue career paths that are congruent with their self-concept and understanding of occupations. In alignment with Gottfredson, Archer et al. (2012) argued that children develop a sense of their science aspirations early in life. In order for science discipline to be a "thinkable" career path for students, the content must be made accessible. When children are not granted access to STEM related content while they are young and curious, they are deterred from a STEM career trajectory and STEM as a career pathway is omitted as an option. Moreover, Young, Feille, and Young (2017) found that African-American females are often deterred from pursuing a science career due to their encounters with dual marginalization and stereotypes early in STEM education. While African-American females possess the necessary competency to be successful in STEM fields, their lack of experience with gender and racially sensitive STEM content results in lack of interest and pursuit of STEM degrees and careers.

If teachers and professional school counselors understand the urban culture where their students are living, they are more apt to connect students' backgrounds and interests with the STEM-related content and occupations. However, teachers and professional school counselors may not know how to do this and need training. Fortunately, the National Science Foundation (NSF) and the leadership team from Advancing Coherent and Equitable Systems of Science Education (ACESSE) project provide editable, open education resource (OER) online training at no cost (Bell, Stromholt, Neill, & Shaw, 2017; see https://docs.google.com/document/d/17hRJU-YXv_e_4w6ySijm4aO4cN70sk_TeVhHr95eCd0/edit).

In addition to training educational professionals, a focus on educating families and offering opportunities to the whole family, especially those with a lower socioeconomic status, should also be a priority. In a 5-year longitudal study, Archer et al. (2017) found that children who had parents and guardians that were exposed to scientific content were more likely to have an interest in STEM. The family plays a pivotal role in shaping a child's perception of what was a viable career path. Cultivating and exposing parent to STEM content enables them to see STEM as a career pathway for their child and provides them with knowledge to cultivate a STEM interest in their children. Thus, schools can offer family STEM nights or STEM career days for the entire family. For example, a school may host a STEM night on a Wednesday evening where parents and students visit learning centers set up around the school to work on problem-based learning activities.

Strategy 2: Plan Unique Urban Experiences to Demonstrate Diversity in STEM Careers

The urban setting can provide unique cultural, economic, and educational opportunities for students. Unfortunately, many of these opportunities are never seized by families. Thus, through field trips to STEM-related sites, career days, and contact with diverse men and women in STEM careers through special guest lectures, teachers and professional school counselors can provide vicarious experiences in STEM to begin to cultivate a positive self-identity in relation to these fields of study. Interacting with and seeing diverse role models who are similar to them (i.e., a Black engineer at a lab or a Hispanic botanist at a conservatory), students' abilities to envision themselves belonging in a STEM career is enhanced and their STEM self-efficacy is improved (Milam, 2012; Syed, Goza, Chemers, & Zurbriggen, 2012).

Strategy 3: Integrate Virtual Reality (VR) Experiences Into the Curriculum to Develop STEM Interest

While the urban setting provides unique opportunities, school budgetary issues can inhibit, at times, providing students with the opportunity to take field trips. Thus, teachers can plan virtual reality field trips for students for whom VR technology is age appropriate. Recent advancements in smartphone and tablet capabilities have provided low-cost options for VR experiences (Zantua, 2017), and the introduction of the Google Cardboard viewfinder with the Expeditions VR system enables teachers to guide students through affordable virtual field trips in the classroom (Brown & Green, 2016; Pierce, 2016). The teacher can use this hardware and software and select field trip locations or persons, which enables students to visit a pharmaceutical biochemistry lab, tour the Great Barrier Reef, or visit with one of NASA's "human computers," Katherine Johnson, an African American woman whose research was instrumental in sending the first American into space. VR opportunities in science positively influence student's STEM interest. In an urban, elementary environment, the use of VR has been shown to expose students to otherwise abstract ideas and vicarious experiences (Isik-Ercan et al., 2014), which, theoretically, can increase STEM self-efficacy. VR experiences can result in students developing an interest in STEM, create a positive self-identity, and help students understand the career opportunities available to women, minorities, and individuals who grew up in urban poverty.

Strategy 4: Develop Science Clubs and Afterschool Science Programs to Help Students See That Individuals Similar to Them Are Interested in STEM

Science programs and clubs do not become prevalent until middle or high school. Almost half of the high schools in the nation utilize science clubs, and only 20% of elementary schools incorporate science programs (Horizon Research, Inc., 2013). Science clubs and afterschool programs are thus needed and underutilized in elementary school, especially in the urban setting. Teachers and professional school counselors have the unique opportunity to create and sponsor these opportunities for elementary students.

After-school programs and clubs are often staffed by older peers and adults from similar demographics and backgrounds as the students being served by the programs; thus, providing students who participate in these programs with role models who are interested in STEM (Coalition for Science After School, 2007; Rhoads, Walden & Winter, 2004). They also provide the opportunity to learn about science, technology, engineering, and math as well as experience success in thinking critically, investigating, and solving problems. Moreover, studies have shown that having these mastery and vicarious experiences in science clubs or afterschool groups can increase students, and especially girls', self-concept, self-efficacy, and confidence as a STEM professional (Wang & Billington, 2016; Levine, Serio, Radaram, Chaudhuri & Talbert, 2016; Gonsalves, Rahm, & Carvalho, 2013).

In developing programs and clubs, it is important to consider the research on the elements that make these experiences effective (in either student achievement, student engagement, or both). Research has shown that effective STEM programs and clubs integrate mastery and vicarious experiences: 1) planning for hands-on, discovery learning opportunities, 2) involving science professionals as guest speakers with a variety of representatives from diverse backgrounds with consideration to but not limited to sex, race, ethnicity, socioeconomic status, and educational levels, 3) seeking out community organization or company personnel (e.g., non-profit groups, local museums, and environmental agencies) to teach students targeted concept/skills and to share with students real-world examples of STEM skills, and 4) gaining support from administration, school board, and students' parents by inviting them to participate and create opportunities to learn alongside the students (Huffling et al., 2017; Blanchard, Hoyle, & Gutierrez, 2017; Flash, Allen, Mack, & Clement, 2017).

Moreover, researchers have demonstrated that stronger bonds and connections are made over longer periods of consistent, continuous, and meaningful experiences (Carpi, Ronan, Falconer, & Lents, 2017). Therefore, programs that are longer in duration are more effective in building students' STEM career self-efficacy.

Strategy 5: Use Inquiry-Based Strategies Where Students Are Invited to Be Participants in Learning That Enable Them to Develop Attitudes and Capabilities Consistent With STEM Professionals

Varelas, Kane, and Wylie (2011) noted that

> *... if we want children, and especially children of color in underserved urban areas, to realize that science is about questioning, testing, measuring, wondering, explaining, and so on, then we need to examine how practices that emphasize obedience to the teacher and encourage seeing the teacher as the only authority in the classroom are consistent or not with such science ideology. If such behavioral codes are in some ways antithetical to the active, inquisitive, questioning, flexible view of science and science learning that teachers try to enact in these classrooms, then how can students make sense of the competing messages they receive? (p. 846)*

To encourage inquisitiveness and questioning, teachers can make simple adaptions to and employ specific instructional strategies during STEM lessons. For example, teachers can encourage students to create variation and adjustments in experiments conducted in a science lab. Teachers can provide choices within STEM projects (Gonsalves, Rahm, & Carvalho, 2013). Allowing students choice when appropriate is an important strategy that invites students to be active participants in their learning, and therefore, creating higher levels of engagement (Beymer, Rosenberg, Schmidt, & Naftzger, 2018; Mortensen & Smart, 2007). These choices available should be relevant to students' urban living situations and their interests to enhance one's self-identity in relation to STEM. For, when lessons are congruent with students' cultural background, motivation has shown to increase (Katz & Assor, 2007). The teacher may also use inquiry-based learning strategies, as they have been shown to provide opportunities for students to connect scientific concepts to their lives and to improve students' perceptions, feelings, and self-identity about STEM (Maxwell et al., 2015; Chen et al., 2014). Self-documentation is another strategy that a teacher or professional school counselor can employ to encourage students to observe, question, and evaluate as a STEM professional would. Self-documentation is a strategy in which students are given digital cameras to take home to document activities in their everyday lives related to specific teacher or counselor provided prompts (Bell, Bricker, Tzou, Lee, & Van Horne, 2012). For example, Bell et al. (2012) stated that a 5th-grade teacher could give her students digital cameras to take pictures of items in the home related to maintaining a healthy lifestyle. Teachers can then ask students to write about the item, how it keeps them from being unhealthy or healthy. Students can then be prompted to share their pictures and writing with peers. Using self-documentation enables students to observe and think about STEM in their everyday lives (Reeve, & Bell, 2009). Additional lesson ideas for integrating self-documentation within STEM curricula could include asking students to:

- Take pictures of biotic and abiotic items in their household and around their home and write about how they engage with these items for an ecosystem lesson.
- Take pictures of clothing they decide to wear to school in the winter and write about the decision making the process for a weather and temperature lesson.
- Take pictures of power kitchen appliances and write about how they interact with renewable and nonrenewable resources for a unit on energy.
- Take pictures of family dinners and categorize the meal into solutions or mixtures for a unit on matter.
- Take pictures of family interactions and reflect on respect and compromise.
- Take pictures of problems that need to be solved and write out possible solutions.

Strategy 6: Enhance Reflective Thinking, Discourse, and Writing Skills to Facilitate Accessibility to STEM Careers and Degrees

Underrepresented races and urban populations in the United States underperform in writing and critical thinking compared to their Asian or Caucasian and suburban counterparts. Writing and critical thinking are crucial components to STEM-related careers and could be a factor in low self-efficacy, deterring these groups from pursuing STEM fields. The most recent National Assessment of Educational Progress' (NAEP) Grade 8 writing assessment reported 44% of Asian and 34% of white students scored proficient or higher compared to 14% Hispanic and 11% Black students in the same category. Also, only 12% of students eligible for free or reduced lunch scored proficient or higher whereas those students not eligible had a 37% passing rate. Students categorized as attending schools in a city scored 23% at proficient or above levels whereas 33% of students in suburban populations earned proficient or higher (National Center for Education Statistics, 2012).

Therefore, to promote STEM career self-efficacy and successful experiences in STEM, elementary teachers, and professional school counselors can provide purposeful exposure to STEM reading, writing, and thinking to students. The Science Writing Heuristic (SWH) approach is one way to create meaningful inquiry experiences and purposefully incorporate scientific discourse and writing (claim, reasoning, evidence, and argumentation) (Cavagnetto, Hand, & Norton-Meier, 2010; Hand, Shelley, Laugerman, Fostvedt, & Therrien, 2018; Hand, Norton-Meier, Gunel, & Akkus, 2016). This "...heuristic is intended to help students construct understanding during practical work. Students are required to produce written explanations of the processes involved in the activity through completion of a template, with particular emphasis placed on claims, evidence, and reflection" (Hand et al. 2002, p. 20). Using SWH during a lesson, a teacher or school counselor may ask students to write about the following in regards to an experiment conducted socially or in a lab: (1) Questions—What are my questions? (2) Experiments—What did I do? (3) Observations—What did I see? (4) Findings/Ideas—What can I say about what I did and saw? What did I discover? (5) Evidence—How do I know this? Why am I saying this about what I did and saw? (6) Contextualization—How do my discoveries compare with other (peers, readings) discoveries? (7) Reflection on Learning—What have I learned? Concept mapping, a method for brainstorming and organizing ideas, is another popular instructional strategy that has been shown to promote reflection and improve writing (Fellows, 1994; Bramwell-Labor, 2014). Journaling is also useful for engaging students in the reflective thinking and writing process (Wist, 2006) and encouraging students in synthesis, argumentation, and critiques ideas (Baxter, Bass, & Glaser, 2001; Minogue, Madden, Bedward, Wiebe, & Carter, 2010; Ruiz-Primo, Li, Shavelson, 2002; Glen & Dotger, 2013).

Strategy 7: Assess and Overtly Confront Stereotypes That May Preclude Students' Interest in STEM Careers

Using Chambers' (1983) Draw a Scientist Test (DAST) or Knight and Cunningham's (2004) Draw an Engineer Test (DAET), a teacher or professional school counselor can quickly assess the stereotypes their students may possess prior to any science or engineering instruction and utilize the outcomes to engage in scientific discourse. For example, if a more significant percentage of girls or underrepresented minorities depict a Caucasian man in their initial drawing, the teacher or counselor can help alleviate this stereotype by visiting or inviting scientists and engineers into the classroom that look like these students. Teachers and professional school counselors can also be mindful of creating mathematical questions or engineering problems geared toward the interests of girls and minorities (Frenzel, 2008; Bowe, Desjardins, Covington Clarkson, & Lawrenz, 2015), for research has shown that using mathematical concepts integrated relevant to girl and minority students improves self-efficacy and challenged stereotypical perceptions about STEM professionals.

Teachers can also create identity safe classrooms (Steele &Cohn-Vargas, 2013) in which students from racial, social, and economic diverse backgrounds in an urban setting can learn and develop STEM identities. Identity-safe classrooms provide students with a safe space to express their thoughts and perceptions without judgement, to engage in rich conversations, and to confront these stereotypes through discussion and reflections (Sparks, 2016); thus, resulting in the opportunity to construct a STEM self-identity and self-efficacy. A detailed description of how to construct an identity safe classroom is beyond the scope of this chapter. More information can be found in Steele and Cohn-Vargas's (2013) text.

Teachers and professional school counselors must also be conscious of their own bias and ensure they give the same opportunities to these students in underrepresented populations in STEM careers and use gender-neutral examples whenever possible. They should be mindful of eliminating stereotypical references or images in classrooms and offices that could influence the interest of girls and minorities in taking STEM courses (Master, Cheryan, & Meltzoff, 2016). For, student's STEM career self-efficacy is shaped by implicit and explicit messages from teachers and other significant individuals. Research has shown that school personnel's gender stereotypes and expectations affect students' STEM attitudes and achievement (Eccles & Jacobs, 1986) and are influential in the loss of girls and minority students in the STEM pipeline (Shapiro & Sax, 2011).

Strategy 8: Promote Self-Insight and Sound Conceptualization About Feasibility of STEM Careers

Professional school counselors are trained and ethically obligated to promote student career development and assist students in developing their skills, interests, and abilities that may improve career decision-making and accessibility (American School Counselor Association, 2012; Holcomb-McCoy, 2007). A professional school counselor, during individual or group sessions, can use goal setting activities, interest inventories, online person-job career assessments, and career websites to promote self-insight and a sound conceptualization of feasible STEM careers. These activities can counteract negative stereotypes and beliefs that girls and individuals from certain social statuses and races cannot be STEM professionals and increase STEM self-efficacy.

Strategy 9: Use Process-Directed Praise and Support to Improve STEM Career Self-Efficacy

Social persuasion is another strategy critical to building the self-efficacy of students, for praise and communication of support from teachers and professional school counselors can profoundly affect the self-efficacy and outcome expectations for STEM coursework and career (Gunderson, Ramirez, Levine, & Beilock, 2012). Researchers state that educators "...should prioritize increasing their [females'] self-efficacy because this seems to be the most important mediator of gender differences in interest in STEM majors" (Tellhed, Bäckström, & Björklund, 2017, p. 93), which can be done through process-oriented praise, emphasizing effort and actions rather than the person. By praising efforts and actions in STEM, professional school counselors and teachers can help reduce the impact of stereotypical gender and racial/ethnic beliefs (Grossman & Porsche, 2013; Blustein et al., 2013). While general praise may reinforce some students' interest in STEM, more effective praise is behavior specific. Behavior-specific praise informs students about the specific behavior they are doing correctly or well. It includes 1) the provision of a positive, verbal statement 2) about an explicit behavior or 3) a specific student or class (Thompson, Marchant, Anderson, Prater, & Gibb, 2012). For example, "Lakesha, I like how you demonstrated to the class the process you used to solve the math problem at the board." Rhode, Jenson, and Reavis (1992) also suggest that effective behavior-specific praise is inclusive IFEED-AV

- Immediately,
- Frequently,
- Enthusiasm,
- Eye Contact,
- Describe,
- Anticipation,
- Variety

A teachers or school counselor's sincerity in provide behavior specific praise is also vital to its effectiveness in reinforcing STEM interest in girls and minorities (Haydon & Musti-Rao, 2011). Table 1 provides a summary of the strategies discussed.

Table 1. Summary of strategies

Strategy	Theory Addressed	Specific Implementation Ideas
Seek Training on Providing Equitable Access by Understanding Students' Cultures and Backgrounds	☑Gottfredson's 1981 Theory of Circumscription and Compromise ☑Bandura's self-efficacy	• Engage in professional development to understand students' cultures and background For example, Advancing Coherent and Equitable Systems of Science Education (ACESSE) project provide editable, open education resource (OER) online training at no cost (Bell, Stromholt, Neill, & Shaw, 2017; see https://docs.google.com/document/d/17hRJU-YXv_e_4w6ySijm4aO4cN70sk_TeVhHr95eCd0/edit) • Involve families in STEM opportunities
Plan Unique Urban Experiences to Demonstrate Diversity in STEM careers.	☑Gottfredson's 1981 Theory of Circumscription and Compromise ☑Bandura's self-efficacy	• Field trips • Guest Speakers • Career Days • Diverse Role Models
Integrate Virtual Reality (VR) Experiences into the Curriculum to Develop STEM Interest.	☑Gottfredson's 1981 Theory of Circumscription and Compromise ☑Bandura's self-efficacy	• Google Cardboard • Virtual Field Trips
Develop Science Clubs and Afterschool Science Programs to Help Students see that Individuals Similar to Them Are Interested in STEM.	☑Gottfredson's 1981 Theory of Circumscription and Compromise ☑Bandura's self-efficacy	• Start elementary STEM clubs • Aspects to incorporate in these programs include: o planning for hands-on, discovery learning opportunities o involving science professionals as guest speakers with a variety of representatives from diverse backgrounds with consideration to but not limited to sex, race, ethnicity, socioeconomic status, and educational levels o seeking out community organization or company personnel (e.g., non-profit groups, local museums, and environmental agencies) to teach students targeted concept/skills and to share with students real-world examples of STEM skill o gaining support from administration, school board, and students' parents by inviting them to participate and create opportunities to learn alongside the students (Huffling et al., 2017; Blanchard, Hoyle, & Gutierrez, 2017; Flash, Allen, Mack, & Clement, 2017).
Use Inquiry-based Strategies where Students Are Invited to be Participants in Learning that Enable Them to Develop Attitudes and Capabilities Consistent with STEM Professionals.	☑Gottfredson's 1981 Theory of Circumscription and Compromise ☑Bandura's self-efficacy	• Encourage students to create variation and adjustments in experiments conducted in a science lab • Provide choices within STEM projects • Provide opportunities for students to connect scientific concepts to their lives • Utilize self-documentation (students are given digital cameras to take home to document activities in their everyday lives related to specific teacher or counselor provided prompts)
Enhance Reflective Thinking, Discourse, and Writing Skills to Facilitate Accessibility to STEM Careers and Degrees.	☑Gottfredson's 1981 Theory of Circumscription and Compromise ☑Bandura's self-efficacy	• Provide purposeful exposure to STEM reading, writing, and thinking to students • Utilize the Science Writing Heuristic (SWH) approach to create meaningful inquiry experiences and purposefully incorporate scientific discourse and writing (claim, reasoning, evidence, and argumentation) (Cavagnetto, Hand, & Norton-Meier, 2010; Hand, Shelley, Laugerman, Fostvedt, & Therrien, 2018; Hand, Norton-Meier, Gunel, & Akkus, 2016) • Incorporate journaling into STEM activities
Assess and Overtly Confront Stereotypes that May Preclude Students' Interest in STEM Careers.	☑Gottfredson's 1981 Theory of Circumscription and Compromise ☑Bandura's self-efficacy	• Use Chambers' (1983) Draw a Scientist Test (DAST) or Knight and/or Cunningham's (2004) Draw an Engineer Test (DAET) • Utilize the identify-safe classroom approach (Steele & Cohn-Vargas, 2013) • Be mindful of eliminating stereotypical references or images in classrooms and offices that could influence the interest of girls and minorities in taking STEM courses (Master, Cheryan, & Meltzoff, 2016).
Promote Self-Insight and Sound Conceptualization about Feasibility of STEM Careers.	☑Gottfredson's 1981 Theory of Circumscription and Compromise ☑Bandura's self-efficacy	Use: • Goal setting activities • Interest inventories • Online person-job career assessments • Career websites
Use Process-directed Praise and Support to Improve STEM Career Self-efficacy.	☑Gottfredson's 1981 Theory of Circumscription and Compromise ☑Bandura's self-efficacy	• Emphasis efforts and actions • Utilize praise • Communicate support

CONCLUSION

The underrepresentation of women and minorities in STEM-related careers continues to be an issue in the United States. According to Bandura's Social Cognitive Theory (1993) and Gottfredson's (2004) Theory of Circumscription and Compromise, decreasing stereotypical perceptions of STEM professionals and increasing self-efficacy in underrepresented populations in STEM at an early age can begin to reverse this perpetual trend. Strategies that have been researched and have shown an improvement in students' STEM self-identity include cultivating scientific interest at an early age by: (1) ensuring that educational professionals understand the cultural background of their students, (2) STEM-related content is accessible to underrepresented populations and is diversity demonstrated in these fields, (3) VR activities are integrated to gain interest, (4) inquiry-based science instruction is included, (5) STEM specific clubs for students are developed, (6) students are engaged in reflective practices involving STEM content, (7). stereotypes are confronted and disproven, (8) promote a positive self-identity and feasibility of STEM careers, and (9) utilizing process-directed praise to support STEM career self-efficacy. While theory and research inform a number of strategies teachers and professional school counselors in an urban setting can employ to improve STEM interest and encourage the pursuit of STEM careers, more of these professionals need to continue to develop and research evidence-based strategies to combat STEM career stereotypes and encourage STEM career aspirations elementary urban school settings (e.g. Taraban, Box, Myers, Pollard, & Bowen, 2007; Weisgram & Bigler, 2007). The lack of empirical research in addition to the need for individuals to pursue STEM careers in the United States, in general, and the underrepresentation of females and minorities in the STEM workforce underscores the urgency of developing empirically supported, holistic models for urban STEM elementary education to improve STEM achievement, STEM interest, and encourage openness to pursue college degrees and careers in STEM fields (National Research Council, 2012; Rivera & Schaefer, 2009). Because learners and societal influences are incredibly complex, the solution may be just as multifaceted as the strategies presented here. One thing that is simple is the understanding that there is more work to be done to create an equitable playing field.

This research received no specific grant from any funding agency in the public, commercial, or not-for-profit sectors.

REFERENCES

American School Counselor Association (ASCA). (2012). *The ASCA National Model: A framework for school counseling programs*. Alexandria, VA: Author. Retrieved from https://www.schoolcounselor.org/asca/media/asca/home/SCCompetencies.pdf

Archer, L., DeWitt, J., Osborne, J., Dillon, J., Willis, B., & Wong, B. (2012). Science aspirations, capital, and family habitus: How families shape children's engagement and identification with science. *American Educational Research Journal*, *49*(5), 881–908. doi:10.3102/0002831211433290

Bandura, A. (1977). Self-efficacy: Toward a unifying theory of behavioral change. *Psychological Review*, *84*(2), 191–215. doi:10.1037/0033-295X.84.2.191 PMID:847061

Bandura, A. (1993). Perceived self-efficacy in cognitive development and functioning. *Educational Psychologist*, *28*(2), 117–128. doi:10.120715326985ep2802_3

Bandura, A. (1997). *Self-efficacy: The exercise of control*. New York, NY: Freeman.

Bandura, A. (2001). Social cognitive theory: An agentic perspective. *Annual Review of Psychology*, *52*(1), 1–26. doi:10.1146/annurev.psych.52.1.1 PMID:11148297

Baxter, G. P., Bass, K. M., & Glaser, R. (2001). Notebook writing in three fifth-grade science classrooms. *The Elementary School Journal*, *102*(2), 123–140. doi:10.1086/499696

Bedard, K., & Cho, I. (2010). Early gender test score gaps across OECD countries. *Economics of Education Review*, *29*(3), 348–363. doi:10.1016/j.econedurev.2009.10.015

Bell, P., Bricker, L., Tzou, C., Lee, T., & Van Horne, K. (2012). Exploring the Science Framework. *Science Scope*, *36*(3), 17–22.

Bell, P., Stromholt, S., Neill, T., & Shaw, S. (2017). *Making science instruction compelling for all students: Using cultural formative assessment to build on learner interest and experience*. Retrieved from http://stemteachingtools.org/pd/SessionC

Beymer, P. N., Rosenberg, J. M., Schmidt, J. A., & Naftzger, N. J. (2018). Examining relationships among choice, affect, and engagement in summer STEM programs. *Journal of Youth and Adolescence*, *47*(6), 1178–1191. doi:10.100710964-018-0814-9 PMID:29356926

Blanchard, C. A., & Lichtenberg, J. W. (2003). Compromise in career decision making: A test of Gottfredson's theory. *Journal of Vocational Behavior, 62*(2), 250–271. doi:10.1016/S0001-8791(02)00026-X

Blanchard, M. R., Hoyle, K. S., & Gutierrez, K. S. (2017). How to start a STEM club. *Science Scope, 41*(3), 88–94. doi:10.2505/4s17_041_03_88

Blustein, D. L., Barnett, M., Mark, S., Depot, M., Lovering, M., Lee, Y., ... DeBay, D. (2013). Examining urban students' constructions of a STEM/career development intervention over time. *Journal of Career Development, 40*(1), 40–67. doi:10.1177/0894845312441680

Bodzin, A., & Gehringer, M. (2001). Breaking science stereotypes. *Science and Children, 35*(1), 36–41.

Bowe, A. G., Desjardins, C. D., Covington Clarkson, L. M., & Lawrenz, F. (2015). Urban elementary single-sex math classrooms: Mitigating stereotype threat for African American girls. *Urban Education, 52*(3), 370–398. doi:10.1177/0042085915574521

Bramwell-Lalor, S., & Rainford, M. (2014). The effects of using concept mapping for improving advanced level biology students' lower- and higher-order cognitive skills. *International Journal of Science Education, 36*(5), 839–864. doi:10.1080/09500693.2013.829255

Brown, A., & Green, T. (2016). Virtual reality: Low-cost tools and resources for the classroom. *TechTrends, 60*(5), 517–519. doi:10.100711528-016-0102-z

Buldu, M. (2006). Young children's perceptions of scientists: A preliminary study. *Educational Research, 48*(1), 121–132. doi:10.1080/00131880500498602

Bureau of Labor Statistics. (2010). *Occupational outlook handbook, 2010-11 Edition, agricultural and food scientists*. Retrieved from http://www.bls.gov/oco/ocos046.htm

Buschor, C. B., Berweger, S., Frei, A. K., & Kappler, C. (2014). Majoring in STEM—What accounts for women's career decision making? A mixed methods study. *The Journal of Educational Research, 107*(3), 167–176. doi:10.1080/00220671.2013.788989

Caleon, I., & Subramaniam, R. (2008). Attitudes towards science of intellectually gifted and mainstream upper primary students in Singapore. *Journal of Research in Science Teaching, 45*(8), 940–954. doi:10.1002/tea.20250

Capobianco, B. M., Diefes-dux, H. A., Mena, I., & Weller, J. (2011). What is an engineer? Implications of elementary school student conceptions for engineering education. *Journal of Engineering Education*, *100*(2), 304–328. doi:10.1002/j.2168-9830.2011.tb00015.x

Carpi, A., Ronan, D. M., Falconer, H. M., & Lents, N. H. (2017). Cultivating minority scientists: Undergraduate research increases self-efficacy and career ambitions for underrepresented students in STEM. *Journal of Research in Science Teaching*, *54*(2), 169–194. doi:10.1002/tea.21341

Cavagnetto, A., Hand, B. M., & Norton-Meier, L. (2010). The nature of elementary student science discourse in the context of the science writing heuristic approach. *International Journal of Science Education*, *32*(4), 427–449. doi:10.1080/09500690802627277

Chambers, D. (1983). Stereotypic images of the scientist: The draw-a-scientist test. *Science Education*, *67*(2), 255–265. doi:10.1002ce.3730670213

Chen, H., Wang, H., Lin, H., Lawrenz, F. P., & Hong, Z. (2014). Longitudinal study of after-school, inquiry-based science intervention on low-achieving students and their perceptions of learning science. *International Journal of Science Education*, *13*(36), 2133–2156. doi:10.1080/09500693.2014.910630

Coalition for Science After School. (2007, March). *Blueprint for Action*. Retrieved from https://www.howtosmile.org/pdf/coalition%20publications/%20Science%20in%20After-School%20blueprint.pdf

Cvencek, D., Meltzoff, A. N., & Greenwald, A. G. (2011). Math-gender stereotypes in elementary school children. *Child Development*, *82*(3), 766–779. doi:10.1111/j.1467-8624.2010.01529.x PMID:21410915

Douglas, K. A., & Strobel, J. (2015). Hopes and goals survey for use in STEM elementary education. *International Journal of Technology and Design Education*, *25*(2), 245–259. doi:10.100710798-014-9277-9

Eccles, J. S., & Jacobs, J. E. (1986). Social forces shape math attitudes and performance. *Signs (Chicago, Ill.)*, *11*(2), 367–380. doi:10.1086/494229

Etzioni, A., & Nunn, C. (1974). The public appreciation of science in contemporary America. *Daedalus*, *103*(3), 191–205.

Fellows, N. J. (1994). A window into thinking: Using student writing to understand conceptual change in science learning. *Journal of Research in Science Teaching, 31*(9), 985–1001. doi:10.1002/tea.3660310911

Finson, K., Pedersen, J., & Thomas, J. (2006). Comparing science teaching styles to students' perceptions of scientists. *School Science and Mathematics, 106*(1), 8–15. doi:10.1111/j.1949-8594.2006.tb18066.x

Flash, K., Allen, M., Mack, T., & Clement, K. (2017). STEM bridges: Evolution of an academic library STEM outreach program. *Journal of Library Administration, 57*(8), 879–890. doi:10.1080/01930826.2017.1374105

Forbes, C. E., & Schmader, T. (2010). Retraining attitudes and stereotypes to affect motivation and cognitive capacity under stereotype threat. *Journal of Personality and Social Psychology, 99*(5), 740–754. doi:10.1037/a0020971 PMID:20822288

Frenzel, M. W. (2008). *Reducing gender stereotypes in mathematics* (Master's Thesis). Retrieved from http://digitalcommons.brockport.edu/ehd_theses/234

Glen, N., & Dotger, S. (2013). Writing like a scientist: Exploring elementary teachers' understandings and practices of writing in science. *Journal of Science Teacher Education, 24*(6), 957–976. doi:10.100710972-013-9348-x

Gonsalves, A., Rahm, J., & Carvalho, A. (2013). 'We could think of things that could be science': Girls' re-figuring of science in an out-of-school-time club. *Journal of Research in Science Teaching, 50*(9), 1068–1097. doi:10.1002/tea.21105

Gottfredson, L. S. (1981). Circumscription and compromise: A developmental theory of occupational aspirations. *Journal of Counseling Psychology, 28*(6), 545–579. doi:10.1037/0022-0167.28.6.545

Gottfredson, L. S. (2002). Gottfredson's theory of circumscription, compromise, and self-creation. In D. Brown (Ed.), *Career Choice and Development* (4th ed.; pp. 85–148). San Francisco: Jossey-Bass.

Gottfredson, L. S. (2004). Applying Gottfredson's theory of circumspection and compromise in career guidance and counseling. In D. Brown & Lent (Eds.), Career Development and Counseling. San Francisco: John Wiley & Sons, Inc.

Gottfredson, L. S., & Lapan, R. (1997). Assessing gender-based circumscription of occupational aspirations. *Journal of Career Assessment, 5*(4), 419–441. doi:10.1177/106907279700500404

Granovskiy, B. (2018). *Science, technology, engineering, and mathematics (STEM) education: An overview*. Washington, DC: Congressional Research Service.

Grossman, J. M., & Porche, M. V. (2013). Perceived gender and racial/ethnic barriers to STEM success. *Urban Education, 49*(6), 698–727. doi:10.1177/0042085913481364

Gunderson, E., Ramirez, G., Levine, S., & Beilock, S. (2012). The role of parents and teachers in the development of gender-related math attitudes. *Sex Roles, 66*(3), 153–166. doi:10.100711199-011-9996-2

Gupta, M. (2004). Enhancing student performance through cooperative learning in physical sciences. *Assessment & Evaluation in Higher Education, 29*(1), 63–73. doi:10.1080/0260293032000158162

Hand, B., Norton-Meier, L., Gunel, M., & Akkus, R. (2016). Aligning teaching to learning: A 3-year study examining the embedding of language and argumentation into elementary science classrooms. *International Journal of Science and Mathematics Education, 14*(5), 847–863. doi:10.100710763-015-9622-9

Hand, B., Shelley, M. C., Laugerman, M., Fostvedt, L., & Therrien, W. (2018). Improving critical thinking growth for disadvantaged groups within elementary school science: A randomized controlled trial using the science writing heuristic approach. *Science Education, 102*(4), 693–710. doi:10.1002ce.21341

Hannah, J. S., & Kahn, S. E. (1989). The relationship of socioeconomic status and gender to the occupational choices of Grade 12 students. *Journal of Vocational Behavior, 34*(2), 161–178. doi:10.1016/0001-8791(89)90012-2

Harwell, M., Moreno, M., Phillips, A., Guzey, S. S., Moore, T. J., & Roehrig, G. H. (2015). A study of STEM assessments in engineering, science, and mathematics for elementary and middle school students. *School Science and Mathematics, 115*(2), 66–74. doi:10.1111sm.12105

Haydon, T., & Musti-Rao, S. (2011). Effective use of behavior-specific praise: A middle school case study. *Beyond Behavior, 20*(2), 31–39.

Helwig, A. (1998). Gender-role stereotyping: Testing theory with a longitudinal sample. *Sex Roles*, *38*(5/6), 403–423. doi:10.1023/A:1018757821850

Holcomb-McCoy, C. (2007). *School counseling to close the achievement gap: A social justice framework for success*. Thousand Oaks, CA: Corwin.

Horizon Research, Inc. (2013). *2012 National survey of science and mathematics education: Highlights Report*. Chapel Hill, NC: Author.

Huffling, L. D., Benavides, A., Ash, M., Germuth, A., Matthews, C., & Carlone, H. B. (2017). Promoting equitable access to STEM in a summer herpetology fieldwork program. (science, technology, engineering and mathematics). *Science Teacher (Normal, Ill.)*, *84*(6), 62. Retrieved from https://search.proquest.com/docview/1933244150

Huguet, P., & Régner, I. (2007). Stereotype threat among schoolgirls in quasi-ordinary classroom circumstances. *Journal of Educational Psychology*, *99*(3), 545–560. doi:10.1037/0022-0663.99.3.545

Isik-Ercan, Z., Zeynep Inan, H., Nowak, J. A., & Kim, B. (2014). 'We put on the glasses and moon comes closer!' urban second graders exploring the earth, the sun and moon through 3D technologies in a science and literacy unit. *International Journal of Science Education*, *36*(1), 129–156. doi:10.1080/09500693.2012.739718

Kanwar, R. (2010). Sustainable water systems for agriculture and 21st century challenges. *Journal of Crop Improvement*, *24*(1), 41–59. doi:10.1080/15427520903307700

Katz, I., & Assor, A. (2007). When choice motivates and when it does not. *Educational Psychology Review*, *19*(4), 429–442. doi:10.100710648-006-9027-y

Knight, M., & Cunningham, C. (2004, June). *Draw an engineer test (DAET): Development of a tool to investigate students' ideas about engineers and engineering*. Paper presented at 2004 American Society for Engineering Education Annual Conference & Exposition, Salt Lake City, UT.

Knight, M., & Cunningham, C. M. (2004). Draw an Engineer Test (DAET): Development of a tool to investigate students' ideas about engineers and engineering. American Society of Engineering Education.

Kramer, D. (2008). Sea perch is a fun, hands-on approach to teaching science. *Physics Today*, *61*(3), 22–23. doi:10.1063/1.2897940

Langdon, D., McKittrick, G., Beede, D., Khan, B., & Doms, M. (2011). *STEM: Good jobs now and for the future*. U.S. Department of Commerce, Economics and Statistics Administration. Retrieved from www.esa.doc.gov/sites/default/files/reports/documents/stemfinalyjuly14_1.pdf

Lent, R. W., Brown, S. D., & Hackett, G. (1994). Toward a unifying social cognitive theory of career and academic interest, choice, and performance. *Journal of Vocational Behavior*, *45*(1), 79–122. doi:10.1006/jvbe.1994.1027

Levine, M., Serio, N., Radaram, B., Chaudhuri, S., & Talbert, W. (2015). Addressing the STEM gender gap by designing and implementing an educational outreach chemistry camp for middle school girls. *Journal of Chemical Education*, *92*(10), 1639–1644. doi:10.1021/ed500945g

Mason, C., Kahle, J., & Gardner, A. (1991). Draw-a-scientist test: Future implications. *School Science and Mathematics*, *91*(5), 193–198. doi:10.1111/j.1949-8594.1991.tb12078.x

Master, A., Cheryan, S., & Meltzoff, A. N. (2016). Computing whether she belongs: Stereotypes undermine girls' interest and sense of belonging in computer science. *Journal of Educational Psychology*, *108*(3), 424–437. doi:10.1037/edu0000061

Master, A., & Meltzoff, A. N. (2016). Building bridges between psychological science and education: Cultural stereotypes, STEM, and equity. *Prospects*, *46*(2), 215–234. doi:10.100711125-017-9391-z

Maxwell, D. O., Lambeth, D. T., & Cox, J. T. (2015). Effects of using inquiry-based learning on science achievement for fifth-grade students. *Asia-Pacific Forum on Science Learning & Teaching*, *1*(16), 106–136.

McKown, C., & Strambler, M. J. (2009). Developmental antecedents and social and academic consequences of stereotype-consciousness in middle childhood. *Child Development*, *80*(6), 1643–1659. doi:10.1111/j.1467-8624.2009.01359.x PMID:19930343

McKown, C., & Weinstein, R. S. (2003). The development and consequences of stereotype consciousness in middle childhood. *Child Development*, *74*(2), 498–515. doi:10.1111/1467-8624.7402012 PMID:12705569

Mead, M., & Metraux, R. (1957). The image of the scientist among high school students: A pilot study. *Science*, *126*(3270), 384–390. doi:10.1126cience.126.3270.384 PMID:17774477

Milam, J. (2012). *Girls and STEM education: A literature review*. Atlanta, GA: Georgia Institute of Technology.

Minogue, J., Madden, L., Bedward, J., Wiebe, E., & Carter, M. (2010). The cross-case analyses of elementary students' engagement in the strands of science proficiency. *Journal of Science Teacher Education*, *21*(5), 559–587. doi:10.100710972-010-9195-y

Morgan, C., Isaac, J., & Sansone, C. (2001). The role of interest in understanding the career choices of female and male college students. *Sex Roles, 44*(5), 295-320. Retrieved from https://search.proquest.com/docview/225379950

Mortensen, M. F., & Smart, K. (2007). Free-choice worksheets increase students' exposure to curriculum during museum visits. *Journal of Research in Science Teaching*, *44*(9), 1389–1414. doi:10.1002/tea.20206

Mortensen, M. F., & Smart, K. (2007). Free-choice worksheets increase students' exposure to curriculum during museum visits. *Journal of Research in Science Teaching*, *44*(9), 1389–1414. doi:10.1002/tea.20206

Museus, S. D., Palmer, R. T., Davis, R. J., & Maramba, D. C. (2011). *Racial and ethnic minority students' success in STEM education. ASHE-Higher Education Report Series*. Hoboken, NJ: Jossey-Bass.

National Academy of Sciences. (2006). *Rising above the gathering storm: Energizing and employing America for a brighter economic future*. Washington, DC: National Academies Press.

National Center for Education Statistics. (2012). *The Nation's Report Card: Writing 2011* (Special Report NCES 2012–470). Institute of Education Sciences, U.S. Department of Education. Retrieved from https://nces.ed.gov/nationsreportcard/pdf/main2011/2012470.pdf

National Research Council. (2012). *A framework for K-12 science education: Practices, crosscutting concepts, and core ideas*. Washington, DC: National Academies Press. Retrieved from https://www.nap.edu/catalog/13165/a-framework-for-k-12-science-education-practices-crosscutting-concepts

National Science Foundation (NSF). (2015). *Women, minorities, and persons with disabilities in science and engineering: 2015* (Special Report NSF 11-309). Arlington, VA: Author. Retrieved from http://www.nsf.gov/2015/nsff15311/tables.cfm

National Science Foundation (NSF), Division of Science Resources Statistics. (2009). *Women, minorities, and persons with disabilities in science and engineering (Special Report NSF 09-305)*. Arlington, VA: Author.

National Science Teachers Association (NSTA). (2003). *Standards for science teacher preparation*. Retrieved from http://static.nsta.org/pdfs/nstastandards2003.pdf

Osborne, J., & Collins, S. (2001). Pupils' views of the role and value of the science curriculum: A focus group study. *International Journal of Science Education*, *23*(5), 441–467. doi:10.1080/09500690010006518

Ozel, M. (2012). Children's images of scientists: Does grade level make a difference? *Educational Sciences: Theory and Practice*, 3187–3198.

Pierce, D. (2016). Inside Google's plan to make VR amazing for absolutely, positively everyone. *Wired.* Retrieved from https://www.wired.com/2016/04/google-vr-clay-bavor/

Reeve, S., & Bell, P. (2009). Children's self-documentation and understanding of the concepts 'healthy' and 'unhealthy.'. *International Journal of Science Education*, *31*(14), 1953–1974. doi:10.1080/09500690802311146

Rhoads, T. R., Walden, S. E., & Winter, B. A. (2002). Sooner elementary engineering and science (SEES) clubs-an informal education experience for elementary children. *Journal for STEM Education*, *5*(3), 47–52. Retrieved from https://ieeexplore.ieee.org/document/1157993

Rhode, G., Jenson, W. R., & Reavis, H. K. (1993). *The tough kid book: Practical classroom management strategies*. Longmont, CO: Sopris West, Inc.

Rivera, L. M., & Schaefer, M. B. (2009). The Career Institute: A collaborative career development program for traditionally underserved secondary (6-12) school students. *Journal of Career Development*, *35*(4), 406–426. doi:10.1177/0894845308327737

Ruiz-Primo, M. A., Li, M., & Shavelson, R. J. (2001). *Looking into students' science notebooks: What teachers do with them? CSE: Technical Report 562*. Los Angeles, CA: Center for Research on Evaluation, Standards, and Student Testing/University of California, Los Angeles.

Satterthwait, D. (2010). Why are 'hands-on' science activities so effective for student learning? *Teaching Science: The Journal of Australian Science Teachers Association*, *56*(2), 7–10.

Scantlebury, K. (2014). Gender matters: Building on the past, recognizing the present, and looking toward the future. In N. G. Lederman & S. K. Abell (Eds.), *Handbook of research on science education* (pp. 187–203). New York: Routledge.

Shapiro, C. A., & Sax, L. J. (2011). Major selection and persistence for women in STEM. *New Directions for Institutional Research*, *152*(152), 5–18. doi:10.1002/ir.404

Shenouda, C. K., & Danovitch, J. H. (2014). Effects of gender stereotypes and stereotype threat on children's performance on a spatial task. *Revue Internationale de Psychologie Sociale*, *27*(3–4), 53–77.

Shepardson, D. P., & Britsch, S. J. (2001). The role of children's journals in elementary school science activities. *Journal of Research in Science Teaching*, *38*(1), 43–69. doi:10.1002/1098-2736(200101)38:1<43::AID-TEA4>3.0.CO;2-I

Shoffner, M. F., Newsome, D., Barrio Minton, C. A., & Wachter Morris, C. A. (2015). A qualitative exploration of the STEM career-related outcome expectations of young adolescents. *Journal of Career Development*, *42*(2), 102–116. doi:10.1177/0894845314544033

Smith, J. L., Sansone, C., & White, P. H. (2007). The stereotyped task engagement process: The role of interest and achievement motivation. *Journal of Educational Psychology*, *99*(1), 99–114. doi:10.1037/0022-0663.99.1.99

Sparks, D. (2016). Reducing stereotype threat in the science and mathematics classroom: An overview of research, best practices, and intervention strategies. *Currents in Teaching and Learning*, *7*(2), 4–17.

Steele, C. M. (1997). A threat in the air: How stereotypes shape intellectual identity and performance. *The American Psychologist*, *52*(6), 613–629. doi:10.1037/0003-066X.52.6.613 PMID:9174398

Steele, D. M., & Cohn-Vargas, B. (2013). *Identity safe classrooms: Places to belong and learn*. Thousand Oaks, CA: Corwin Press.

Swanson, J. L., & Fouad, N. A. (2014). *Career theory and practice: Learning through case studies* (3rd ed.). Thousand Oaks, CA: Sage.

Syed, M., Goza, B. K., Chemers, M. M., & Zurbriggen, E. L. (2012). Individual differences in preferences for matched-ethnic mentors among high-achieving ethnically diverse adolescents in STEM. *Child Development, 83*(3), 896–910. doi:10.1111/j.1467-8624.2012.01744.x PMID:22506789

Taraban, R., Box, C., Myers, R., Pollard, R., & Bowen, C. (2007). Effects of active-learning experiences on achievement, attitudes, and behaviors in high school biology. *Journal of Research in Science Teaching, 44*(7), 960–979. doi:10.1002/tea.20183

Tellhed, U., Bäckström, M., & Björklund, F. (2017). Will I fit in and do well? The importance of social belongingness and self-efficacy for explaining gender differences in interest in STEM and HEED majors. *Sex Roles, 77*(1), 86–96. doi:10.100711199-016-0694-y PMID:28725103

Thompson, M. T., Marchant, M., Anderson, D., Prater, M. A., & Gibb, G. (2012). Effects of tiered training on general educators' use of specific praise. *Education & Treatment of Children, 35*(4), 521–546. doi:10.1353/etc.2012.0032

Thurston, A., Topping, K., Tolmie, A., Christie, D., Karagiannidou, E., & Murray, P. (2010). Cooperative learning in science: Follow-up from primary to high school. *International Journal of Science Education, 32*(4), 501–522. doi:10.1080/09500690902721673

Türkmen, H. (2008). Turkish primary students' perceptions about scientists and what factors affecting the image of scientists. *Eurasia Journal of Mathematics, Science &Technology Education, 4*(1), 55–61.

United States Department of Labor (USDOL). (2013). *Employment projections: 2012-2022 summary*. Washington, DC: Author. Retrieved from http://www.bls.gov/news.release/ecopro.nr0.htm

U.S. Census Bureau. (2012, April 8). *Recent population trends for the U.S. island areas: 2000 to 2010* (Report No. P23-213). Washington, DC: U.S. Government Printing Office.

Varelas, M., Kane, J. M., & Wylie, C. D. (2011). Young African American children's representations of self, science, and school: Making sense of difference. *Science Education, 95*(5), 824–851. doi:10.1002ce.20447

Wang, H., & Billington, B. (2016). Economically disadvantaged minority girls' knowledge and perceptions of science and engineering and related careers. *Journal of Extension, 54*(6), 1–8.

Weisgram, E. S., & Bigler, R. S. (2007). Effects of learning about gender discrimination on adolescent girls' attitudes toward and interest in science. *Psychology of Women Quarterly, 31*(3), 262–269. doi:10.1111/j.1471-6402.2007.00369.x

Wist, C. (2006). *Putting it all together: Understanding the research behind interactive notebooks.* Retrieved from https://tccl.arcc.albany.edu/knilt/images/d/d0/Interactive_Notebooks_Research.pdf

Young, J. L., Feille, K. K., & Young, J. R. (2017). Black girls as learners and doers of science: A single-group summary of elementary science achievement. *Electronic Journal of Science Education, 21*(2).

Zantua, L. (2017). Utilization of virtual reality content in grade 6 social studies using affordable virtual reality technology. *Asia Pacific Journal of Multidisciplinary Research, 5*(2), 1-10. Retrieved from https://doaj.org/article/2fe23fbd126948b08cffc832a5018a7

KEY TERMS AND DEFINITIONS

Stereotype: Oversimplified, yet fixed, idea or perception.

Conclusion

As noted throughout this book, efforts are being made to enhance science, technology, engineering, and mathematics (STEM) education for urban K-12 students across the nation. Many of these efforts have proved fruitful, yet many others have brought to light additional challenges, unintended consequences, and persisting barriers that still require close examination and very intentional efforts to eradicate. When considering the research literature surrounding K-12 urban STEM learning, three main areas of concern are noted: poverty and access to STEM resources, unequal participation among women, and unequal participation among minorities. These areas of concern are not unique to K-12 learning environments, as they pervade post-secondary education and the STEM workforce as well. However, they can and should be addressed early on.

We know that the United States must bolster the number of individuals entering the STEM workforce (AAUW, 2016; Beckett, Maltbie, Wright, Sherman, & Sersion, 2016; NSF, 2017). This includes allowing for participation of a diverse workforce and making use of talent that, if ignored, may have profound impacts on society. We know that we must make accessible opportunities for engagement in STEM for "Black, Brown and poor children" (Howard & Milner, 2014, p. 201), as we note in the Preface. We are well aware as a nation that an equity disparity continues to plague STEM education and career fields in the US (AAUW, 2016; NSF, 2017). Yet, despite efforts to improve the state of STEM education and the demographic characteristics of the STEM workforce, we are still falling short. Those interested in supporting equitable access, engagement, and interest in STEM fields proclaim the need to consider social transformation and development of environments that support the inclusion of individuals from all genders and races.

Research has begun to examine issues of social transformation and social justice that surround the implementation of STEM in urban areas. It is not uncommon for schools in urban areas to be labeled as "failing" due to low graduation rates and low testing scores, among other factors. These schools are faced with the daunting task of miraculously increasing test scores and graduation rates, often with even fewer resources, or to be closed. Those that do not rise to appropriate levels are shuttered. In their place, STEM-intensive schools are often created (Bullock, 2017). On first glance, this replacement can be beneficial to the surrounding community as students who attend receive skill-specific training in order to matriculate into the STEM workforce, thus contributing to sustaining the STEM infrastructure of the surrounding community. However, these STEM-intensive schools often practice selective submission processes, which may result in attendee demographics that are substantially different from the surrounding community (Bullock, 2017; Nasir & Vakil, 2017).

Nasir & Vakil (2017) examined STEM academies in urban areas in California and noted that, in many cases, the population of students admitted to such academies were largely White even when the surrounding community was largely Black and/or Latino. This raises questions regarding why those underrepresented in STEM—those for whom many of the academies were created—are not equitably represented. Furthermore, selective admissions processes may also result in some groups, in most instances Black and Brown students, being transported outside of their communities to schools in other areas. These practices in turn serve to strengthen racial and economic stratification, thus amounting to practices that oppress and restrict those underrepresented in STEM (Bullock, 2017).

Further examination has shown that even when Black and Latino students are admitted to meritocratic schools, they experience a culture that is more exclusive than inclusive and is apt to perpetuate microaggressions (Nasir & Vakil, 2017) that impact students' ability to engage in the learning experience, the learning community, and thus, their ability to experience persistence and success. Thus, the question raised is whether the reason behind the inequity in representation among gender and race in STEM fields is due to lack of interest in STEM fields or rather is a result of an environment hostile to minorities, women, or those from low income families. Some Black and Brown urban high school students have reported their "experiences within [STEM] academies to be incongruent with their racial and gender identities" (Nasir & Vakil, 2017, p. 386). Other students indicated that "the most rigorous academic spaces…were not easily accessible to Black and Brown students and that the few high-achieving students of color or girls who did gain acceptance to these academies did not feel supported" (Nasir & Vakil, 2017, p. 387). Thus, it appears that some practices among STEM-centered schools serving urban and minority populations may be more aligned with power, privilege, and marginalization than inclusivity and equitable representation.

Another issue noted within the literature is the lack of clear guidance on what a STEM school or program should look like—the structure, curriculum, and nuances of implementation. As Bullock (2017) describes, "absent clear criteria for what makes a successful STEM school beyond investments in state-of-the-art technology and quality teachers, school districts are left to develop their own plans" (p. 629). Thus, the question is then raised regarding whether STEM schools and programs inherently serve those who are underrepresented in such fields and, if so, to what extent, how, and in what specific circumstances.

Some states, for instance, engage in the practice of formal STEM certification for schools and programs in order to bolster STEM participation and to provide opportunities for STEM engagement in early adolescence. Surprisingly, little research has considered whether these initiatives have yielded the intended outcomes. One study examined whether there were differences in minority student achievement when attending a STEM certified school compared to a non-STEM certified school in an urban area of Georgia (Wendt, Rockinson-Szapkiw, & Cordes, 2017). The findings suggested that third grade minority students who were enrolled in a STEM certified school experienced higher levels of science achievement than those who were not enrolled in a STEM certified school. However, at the fifth grade level, minority students who were enrolled in a STEM certified school demonstrated lower levels of science achievement than those who were not enrolled in a STEM certified school. Thus, it is unclear whether participation in STEM certified schools is of benefit to Black and Brown students in urban learning environments. While the study was limited in scope, it raises questions regarding the nuances of how STEM learning is implemented within schools, including teacher training and practices, as well as specific practices in place for supporting a minority population.

Another study examined implementation of STEM-centered initiatives, including the Peer Enabled Restructured Classroom (PERC) model, Project Lead the Way (PLTW), and the recently adopted *Next Generation Science Standards* among charter schools in one school district in St. Louis (Icel, 2018). Results indicated a need to not only focus on STEM policy, but also the nuances of how such policy is enacted at the district, school, and classroom levels. Specifically, the study shed light on the importance of stakeholder motivation, recruiting and supporting full-time and long term school leaders for capacity building, flexibility in attitudes and willingness to adopt change, and reflective action. Staff motivation, administrative support, and professional development were cited as paramount to ensuring the success of STEM implementation in the K-12 classroom (Icel, 2018).

Research examining teacher preparedness for implementing STEM curriculum has supported the need for professional development (PD). However, mixed results have been found regarding the efficacy of such PD, especially related to student learning outcomes (Capraro et al., 2016). Studies show, however, that PD that enacts significant change in student achievement should be high quality, intensive, and sustained (Capraro et al., 2016; Yoon et al., 2007).

Some research has focused on out of school time for implementation of STEM-centered initiatives in order to increase student engagement in STEM fields through authentic experiences. For example, Djonko-Moore et al. (2018) implemented an experiential science education experience among 3rd-6th grade Black and Latino children from urban communities in Colorado in an effort to enhance content knowledge and engagement in science. The findings suggested that students who participated experienced enhanced science content knowledge and were more engaged in science learning. Thus, the authors concluded that implementing culturally relevant experiential learning opportunities are essential for encouraging student interest in and engagement in learning.

While limited research has explored the efficacy of STEM integration and how best to prepare teachers for urban STEM education, there are evidence-based practical strategies that have yielded positive results. For instance, Beckett et al.'s (2016) examination of the implementation of CincySTEM iTEST project-based learning, in conjunction with providing access to technological devices for both in-school and out-of-school use, among low-income Black high school students has shown that students' engagement in science, engineering, and the overall learning process is positively affected by units of study that utilize interactive and hands-on activity, collaboration with peers and teams, direct linkages to neighborhood and community, in-school and out-of-school learning opportunities, and meaningful experiences.

As K-12 schools seek to integrate STEM in their current or revised curriculum, the literature supports carefully designed and constructed learning opportunities that follow evidence-based approaches, intentional and sustainable training and support of expert teachers, partnership with community stakeholders, intentional alignment with real-world situations, problems, and experiences (Becket et al., 2016).

In this book, the chapter authors have addressed these challenges, as well as social transformation and the development of effective and equitable learning environments. Given the complexity of the issues that arrive hand-in-hand with STEM initiatives in urban communities, there is still much research that remains. The overview of selected literature provided here and the chapters that our collective authors have constructed for this book provide a foundation for more accurate and in-depth understanding of the current state of K-12 STEM education in urban learning environments, efforts to prepare and provide professional development to K-12 educators, and practical strategies for enacting change among access to, opportunity in, and support through STEM education. As the nation moves forward in supporting STEM education for all students, we must consider the barriers that persist in supporting and retaining a diverse STEM workforce. This begins at the earliest stages of education, and is the responsibility of parents, care-takers, teachers, administrators, policy makers, and all citizens as stakeholders in our youth's future. It is our hope that this book will serve as a catalyst for engaging all in acting as agents of change and making real and meaningful strides towards ensuring equity and quality in urban K-12 STEM education.

Jillian L. Wendt
University of the District of Columbia, USA

REFERENCES

American Association for University Women (AAUW). (2016). *Barriers and bias: The status of women in leadership*. Washington, DC: AAUW.

Beckett, G. H., Hemmings, A., Maltbie, C., Wright, K., Sherman, M., & Sersion, B. (2016). Urban high school student engagement through CincySTEM iTEST projects. *Journal of Science Education and Technology*, *25*(6), 995–1007. doi:10.100710956-016-9640-6

Bullock, E. C. (2017). Only STEM can save us? Examining race, place, and STEM education as property. *Educational Studies*, *53*(6), 628–641. doi:10.1080/00131946.2017.1369082

Capraro, R. M., Capraro, M. M., Scheurich, J. J., Jones, M., Morgan, J., Huggins, K. S., ... Han, S. (2016). Impact of sustained professional development in STEM on outcome measures in a diverse urban district. *The Journal of Educational Research*, *109*(2), 181–196. doi:10.1080/00220671.2014.936997

Djonko-Moore, C. M., Leonard, J., Holifield, Q., Bailey, E. B., & Almughyirah, S. M. (2018). Using culturally relevant experiential education to enhance urban children's knowledge and engagement in science. *Journal of Experiential Education*, *41*(2), 137–153. doi:10.1177/1053825917742164

Howard, T. C., & Milner, H. R. (2014). Teacher preparation for urban schools. In R. Milner & K. Lomotey (Eds.), *Handbook of urban education* (pp. 199–216). New York, NY: Routledge.

Icel, M. (2018). Implementation of STEM policy: A case study of a STEM-focused urban charter school. *Journal of STEM Education*, *19*(3), 7–13.

Nasir, N. S., & Vakil, S. (2017). STEM-focused academies in urban schools: Tensions and possibilities. *Journal of the Learning Sciences*, *26*(3), 376–406. doi:10.1080/10508406.2017.1314215

National Science Foundation. (2017). *Women, minorities, and persons with disabilities in science and engineering*. Retrieved from https://www.nsf.gov/statistics/2017/nsf17310/

Wendt, J. L., Rockinson-Szapkiw, A. J., & Cordes, M. (2017). Examining the influence of a STEM certification model on female, minority science outcomes. *The Journal of Educational Research*, *111*(6), 666–677. doi:10.1080/00220671.2017.1396437

Yoon, K. S., Duncan, T. W.-Y., Lee, S., Scarloss, B., & Shapley, K. L. (2007). *Reviewing the evidence on how teacher professional development affects student achievement*. U.S. Department of Education, Institute of Education Sciences. National Center for Education Evaluation and Regional Assistance, Regional Educational Laboratory Southwest. Retrieved from http://ies.ed.gov/ncee/edlabs/regions/southwest/pdf/rel_2007033_sum.pdf

Glossary

Academic Language: The language that helps students engage effectively with academic content.

African American English (AAE): A variety of English spoken by many African Americans in the United States. Also referred to as African American Vernacular English, Black English, and Ebonics.

Attitude: Attitude toward STEM subjects describes a participant's favor or disfavor of science, technology, engineering, or mathematics as an academic pursuit.

Conservative Protestants: Individuals comprising a religious subgroup believing the Bible to be the ultimate source for truth claims about life and holding a literalist viewpoint of the origins of life as found in the book of Genesis.

Critical Language Awareness: The idea that language conventions and language practices are invested with power relations and ideological processes which people are not aware of.

District Initiatives: Plans or strategies implemented at the district level that impact teaching and learning within schools.

English Learner (EL): A student who is in the process of acquiring English and speaks a language other than English at home. Also knows as emergent bilinguals (EBs) or English language learners (ELLs).

Integrated Curriculum: A curriculum that incorporates literacy and math standards within the STEM curriculum.

Interest: The combination of emotional attachment and personal valuation of a task resulting in a desire for various levels of enjoyment.

Linguistics: The scientific study of language, its structure, and use.

Microinsults: Communications that demean a person's identity.

Microinvalidations: Nullification or negation of thoughts, feelings, and realities of specific groups of people.

Pedagogical Modeling: Content and/or materials that provide students with a model of the desired instructional approaches.

Principal Leadership: Actions of the building leader that provide guidance and direction to improve teaching and learning within a school.

Professional Development (PD): Training, formal education, or advanced professional learning intended to help administrators, teachers, and other educators improve their professional knowledge, competence, skill, and effectiveness.

Professional Development: Opportunities for teachers and staff to engage in learning specific to topics related to improving their curriculum and delivery of instruction.

Professional Learning Communities: Groups of teachers working together to improve teaching and learning within their classrooms.

STEM Career Interest: A participant's desire to pursue a career in a STEM field due to a perception of potential enjoyment.

STEM Curriculum: Curriculum integrating science, technology, engineering, and math teaching and learning standards.

STEM Professional: An individual who works in a science, engineering, technology or mathematical field and who holds a minimum of a Bachelor's degree in their professional field.

Stereotype: Oversimplified, yet fixed, idea or perception.

Teacher Agency: A teacher's willingness and ability to take action and make changes.

Teacher Leadership: Ability of teachers to influence teaching and learning beyond their own classrooms.

Urban Education: Schools in large, densely populated areas with diverse populations, typically characterized by large enrollments and complex bureaucratic systems.

Compilation of References

Abd-El-Khalick, F., Boujaoude, S., Duschl, R., Lederman, N. G., Mamlok-Naaman, R., Hofstein, A., ... Tuan, H. L. (2004). Inquiry in science education: International perspectives. *Science Education*, *88*(3), 397–419. doi:10.1002ce.10118

About the Standards. (n.d.). Retrieved 5 29, 2018, from Common Core State Standards Initiative: http://www.corestandards.org/about-the-standards

Adamuti-Trache, M., & Andres, L. (2008). Embarking on and persisting in scientific fields of study: Cultural capital, gender, and curriculum along the science pipeline. *International Journal of Science Education*, *30*(12), 1557–1584. doi:10.1080/09500690701324208

Aguilar, E. (2013). *The art of coaching: Effective strategies for school transformation*. Hoboken, NJ: John Wiley & Sons.

Aikenhead, G. S. (1996). Science education: Border crossing into the subculture of science. *Studies in Science Education*, *27*(1), 1–52. doi:10.1080/03057269608560077

Ajzen, I. (1985). From intentions to actions: A theory of planned behavior. In J. Kuhl & J. Beckman (Eds.), *Action-control: From cognition to behavior* (pp. 11–39). Heidelberg, Germany: Springer-Verlag. doi:10.1007/978-3-642-69746-3_2

Ajzen, I., & Fishbein, M. (1980). *Understanding attitudes and predicting behavior*. Englewood Cliffs, NJ: Prentice Hall.

Alfouaim, A. (2012). *The critical language awareness impact on language variation*. California State University.

Alim, S. H. (2005). Critical language awareness in the United States: Revisiting issues and revisiting pedagogies in a resegregated society. *Educational Researcher*, *34*(7), 24–31. doi:10.3102/0013189X034007024

Allinace, A. (2014). American After PM: Afterschool Programs in Demand. Washington, DC: Academic Press.

Altinyelken, H. K. (2015). *Evolution of Curriculum Systems to Improve Learning Outcomes and Reduce Disparities in School Achievement.* Academic Press.

American Association for the Advancement of Science (AAAS). (1989). Project 2061: Science for All Americans. Washington, DC: AAAS.

American Association for the Advancement of Science (AAAS). (1993). *Benchmarks for Science Literacy*. New York: AAAS.

American Association for the Advancement of Science (AAAS). (2001a). *Atlas of Science Literacy* (Vol. 1-2). Washington, DC: American Association for the Advancement of Science.

American Association for the Advancement of Science. (2015). *Vision and change in undergraduate biology education: A call to action.* Retrieved from http://visionandchange.org/files/2013/11/aaas-VISchange-web1113.pdf

American Association for University Women (AAUW). (2016). *Barriers and bias: The status of women in leadership*. Washington, DC: AAUW.

American School Counselor Association (ASCA). (2012). *The ASCA National Model: A framework for school counseling programs.* Alexandria, VA: Author. Retrieved from https://www.schoolcounselor.org/asca/media/asca/home/SCCompetencies.pdf

Anderson, L. e. (2013). *A Taxonomy for Learning, Teaching, and Assessing: A Revision of Bloom's Taxonomy of Educational Objectives, Abridged Edition*. Pearson.

Anderson, R. D. (2002). Reforming science teaching: What research says about inquiry. *Journal of Science Teacher Education, 13*(1), 1–12. doi:10.1023/A:1015171124982

Annetta, L. A., Frazier, W. M., Folta, E., Holmes, S., Lamb, R., & Cheng, M. T. (2013). Science teacher efficacy and extrinsic factors toward professional development using video games in a design-based research model: The next generation of STEM learning. *Journal of Science Education and Technology, 22*(1), 47–61. doi:10.100710956-012-9375-y

Apugo, D., Wendt, J. L., Deksissa, T., & Peters, M. (2018). The science behind support: An informed perspective on preparing urban science teachers for diverse 21st century classrooms. In A. S. Burtin, J. S. Fleming, & P. Hampton-Garland (Eds.), Changing urban landscapes through public higher education (pp. 97-115). Academic Press.

Archer, L., Dewitt, J., & Osborne, J. (2015). Is Science for Us? Black Students' and Parents' Views of Science and Science Careers. *Science Education*, *99*(2), 199–237. doi:10.1002ce.21146 PMID:28579645

Archer, L., Dewitt, J., Osborne, J., Dillon, J., Willis, B., & Wong, B. (2010). "Doing" science versus "being" a scientist: Examining 10/11_year_old schoolchildren's constructions of science through the lens of identity. *Science Education*, *94*(4), 617–639. doi:10.1002ce.20399

Archer, L., DeWitt, J., Osborne, J., Dillon, J., Willis, B., & Wong, B. (2012). "Balancing acts": Elementary school girls' negotiations of femininity, achievement, and science. *Science Education*, *96*(6), 967–989. doi:10.1002ce.21031

Archer, L., DeWitt, J., Osborne, J., Dillon, J., Willis, B., & Wong, B. (2012). Science aspirations, capital, and family habitus: How families shape children's engagement and identification with science. *American Educational Research Journal*, *49*(5), 881–908. doi:10.3102/0002831211433290

Archer, L., DeWitt, J., Osborne, J., Dillon, J., Willis, B., & Wong, B. (2013). 'Not girly, not sexy, not glamorous': Primary school girls' and parents' constructions of science aspirations. *Pedagogy, Culture & Society*, *21*(1), 171–194. doi:10.1080/14681366.2012.748676

Arifeen, S. R., & Gatrell, C. (2013). A blind spot in organization studies: Gender with ethnicity, nationality and religion. *Gender in Management*, *28*(3), 151–170. doi:10.1108/GM-01-2013-0008

Artiles, A. J., Harry, B., Reschly, D. J., & Chinn, P. C. (2002). Over-identification of students of color in special education: A critical overview. *Multicultural Perspectives*, *4*(1), 3–10. doi:10.1207/S15327892MCP0401_2

Ashley, M., Cooper, K. M., Cala, J. M., & Brownell, S. (2017). Building Better Bridges into STEM: A Synthesis of 25 Years of Literature on STEM Summer Bridge Programs. *CBE Life Sciences Education*, *16*(4), e3. doi:10.1187/cbe.17-05-0085 PMID:29146667

Astley, J., & Francis, L. J. (2010). Promoting positive attitudes towards science and religion among sixth-form pupils: Dealing with Scientism and Creationism. *British Journal of Religious Education*, *32*(3), 189–200. doi:10.1080/01416200.2010.498604

August, D., Artzi, L., & Mazrum, J. (2010). *Improving science and vocabulary learning of English language learners (CREATE Brief)*. Washington, DC: Center for Research on the Educational Achievement and Teaching of English Language Learners.

Ayalon, H., & Livneh, I. (2013). Educational standardization and gender differences in mathematics achievement: A comparative study. *Social Science Research, 42*(2), 432–445. doi:10.1016/j.ssresearch.2012.10.001 PMID:23347486

Bair, A. N., & Steele, J. R. (2010). Examining the consequences of exposure to racism for the executive functioning of Black students. *Journal of Experimental Social Psychology*, *46*(1), 127–132. doi:10.1016/j.jesp.2009.08.016

Baker, S. (2014). Faith-based schools and the creationism controversy: The importance of the meta-narrative. In *International handbook of learning, teaching and leading in faith-based schools* (pp. 177–192). New York, NY: Springer Netherlands. doi:10.1007/978-94-017-8972-1_9

Bandura, A. (1977). Self-efficacy: Toward a unifying theory of behavioral change. *Psychological Review*, *84*(2), 191–215. doi:10.1037/0033-295X.84.2.191 PMID:847061

Bandura, A. (1977). *Social learning theory*. New York, NY: General Learning Press.

Bandura, A. (1986). *Social foundations of thought and action: A social cognitive theory*. Englewood Cliffs, NJ: Prentice Hall.

Bandura, A. (1993). Perceived self-efficacy in cognitive development and functioning. *Educational Psychologist*, *28*(2), 117–128. doi:10.120715326985ep2802_3

Bandura, A. (1997). *Self-efficacy: The exercise of control*. New York: W.H. Freeman.

Bandura, A. (2001). Social cognitive theory: An agentic perspective. *Annual Review of Psychology*, *52*(1), 1–26. doi:10.1146/annurev.psych.52.1.1 PMID:11148297

Banerjee, P. A., & Lamb, S. (2016). A systematic review of factors linked to poor academic performance of disadvantaged students in science and maths in schools. *Cogent Education, 3*(1).

Banilower, E. R., Heck, D. J., & Weiss, I. R. (2007). Can professional development make the vision of the standards a reality? The impact of the national science foundation's local systemic change through teacher enhancement initiative. *Journal of Research in Science Teaching*, *44*(3), 375–395. doi:10.1002/tea.20145

Banton, M. (2011). Religion, faith, and intersectionality. *Ethnic and Racial Studies*, *34*(7), 1248–1253. doi:10.1080/01419870.2011.582727

Bardick, A., Bernes, K., Magnusson, K., & Witko, K. (2008). Junior high school students' career plans for the future. *Journal of Career Development*, *32*(3), 250–271. doi:10.1177/0894845305279168

Barkana, R. (2018). Possible interaction between baryons and dark-matter particles revealed by the first stars. *Nature*, *555*(7694), 71–74. doi:10.1038/nature25791 PMID:29493590

Barmby, P., Kind, P., & Jones, K. (2008). Examining changing attitudes in secondary school science. *International Journal of Science Education*, *30*(8), 1075–1093. doi:10.1080/09500690701344966

Baxter, G. P., Bass, K. M., & Glaser, R. (2001). Notebook writing in three fifth-grade science classrooms. *The Elementary School Journal*, *102*(2), 123–140. doi:10.1086/499696

Beckett, G. H., Hemmings, A., Maltbie, C., Wright, K., Sherman, M., & Sersion, B. (2016). Urban high school student engagement through CincySTEM iTEST projects. *Journal of Science Education and Technology*, *25*(6), 995–1007. doi:10.100710956-016-9640-6

Bedard, K., & Cho, I. (2010). Early gender test score gaps across OECD countries. *Economics of Education Review*, *29*(3), 348–363. doi:10.1016/j.econedurev.2009.10.015

Bell, P., Stromholt, S., Neill, T., & Shaw, S. (2017). *Making science instruction compelling for all students: Using cultural formative assessment to build on learner interest and experience.* Retrieved from http://stemteachingtools.org/pd/SessionC

Bell, P., Bricker, L., Tzou, C., Lee, T., & Van Horne, K. (2012). Exploring the Science Framework. *Science Scope*, *36*(3), 17–22.

Bem, S. L. (1977). On the utility of alternative procedures for assessing psychological androgyny. *Journal of Consulting and Clinical Psychology*, *45*(2), 196–205. doi:10.1037/0022-006X.45.2.196 PMID:850004

Bem, S. L. (1981). Gender schema theory: A cognitive account of sex typing. *Psychological Review*, *88*(4), 354–364. doi:10.1037/0033-295X.88.4.354

Bem, S. L. (1983). Gender schema theory and its implications for child development: Raising gender-aschematic children in a gender-schematic society. *Signs (Chicago, Ill.)*, *8*(4), 598–616. doi:10.1086/493998

Bem, S. L. (1984). Androgyny and gender schema theory: A conceptual and empirical integration. In R.A. Dienstbier, & T.B. Sonderegger (Eds.), *Nebraska Symposium on Motivation* (Vol. 34). Lincoln, NE: University of Nebraska Press.

Bem, S. L. (1993). *The lenses of gender: Transforming the debate on sexual inequality*. New Haven, CT: Yale University Press.

Benbow, C. (2012). Identifying and nurturing future innovators in science, technology, engineering, and mathematics: A review of findings from the Study of Mathematically Precocious Youth. *Peabody Journal of Education*, *87*(1), 16–25. doi:10.1080/0161956X.2012.642236

Bennett, A., Bridglall, B., Cauce, A., Everson, H., Gordon, E., Lee, C., ... Stewart, J. (2004). *All Students Reaching the Top: Strategies for Closing Academic Achievement Gaps. A Report of the National Study Group for the Affirmative Development of Academic Ability*. Naperville, IL: Learning Point Associates.

Bennett, J., & Hogarth, S. (2009). Would you want to talk to a scientist at a party? High school students' attitudes to school science and to science. *International Journal of Science Education*, *31*(14), 1975–1998. doi:10.1080/09500690802425581

Best, R. K., Edelman, L. B., Krieger, L. H., & Eliason, S. R. (2011). Multiple disadvantages: An empirical test of intersectionality theory in EEO litigation. *Law & Society Review*, *45*(4), 991–1025. doi:10.1111/j.1540-5893.2011.00463.x

Betz, D. E., & Sekaquaptewa, D. (2012). My fair physicist? Feminine math and science role models demotivate young girls. *Social Psychological & Personality Science*, *3*(6), 738–746. doi:10.1177/1948550612440735

Beymer, P. N., Rosenberg, J. M., Schmidt, J. A., & Naftzger, N. J. (2018). Examining relationships among choice, affect, and engagement in summer STEM programs. *Journal of Youth and Adolescence*, *47*(6), 1178–1191. doi:10.100710964-018-0814-9 PMID:29356926

Billingsley, B. (2013). Students' perceptions of apparent contradictions between science and religion: Creation is only the beginning. In N. Mansour & R. Wegerif (Eds.), *Science education for diversity: Theory and practice* (pp. 329–338). New York, NY: Springer Netherlands. doi:10.1007/978-94-007-4563-6_16

Billingsley, B., Taber, K., Riga, F., & Newdick, H. (2013). Secondary school students' epistemic insight into the relationships between science and religion—A preliminary enquiry. *Research in Science Education*, *43*(4), 1715–1732. doi:10.100711165-012-9317-y

Bishop, R., Berryman, M., Cavanagh, T., & Teddy, L. (2009). Te Kotahitanga: Addressing educational disparities facing Māori students in New Zealand. *Teaching and Teacher Education*, *25*(5), 734–742. doi:10.1016/j.tate.2009.01.009

Blanchard, C. A., & Lichtenberg, J. W. (2003). Compromise in career decision making: A test of Gottfredson's theory. *Journal of Vocational Behavior*, *62*(2), 250–271. doi:10.1016/S0001-8791(02)00026-X

Blanchard, M. R., Hoyle, K. S., & Gutierrez, K. S. (2017). How to start a STEM club. *Science Scope*, *41*(3), 88–94. doi:10.2505/4s17_041_03_88

Blustein, D. L., Barnett, M., Mark, S., Depot, M., Lovering, M., Lee, Y., ... DeBay, D. (2013). Examining urban students' constructions of a STEM/career development intervention over time. *Journal of Career Development*, *40*(1), 40–67. doi:10.1177/0894845312441680

Boaler, J. (2016). *Mathematical mindsets: Unleashing students' potential through creative math, inspiring messages and innovative teaching*. John Wiley & Sons.

Bodzin, A., & Gehringer, M. (2001). Breaking science stereotypes. *Science and Children*, *35*(1), 36–41.

Bøe, M. V., Henriksen, E. K., Lyons, T., & Schreiner, C. (2011). Participation in science and technology: Young people's achievement-related choices in late-modern societies. *Studies in Science Education*, *47*(1), 37–72. doi:10.1080/03057267.2011.549621

Borko, H. (2004). Professional development and teacher learning: Mapping the terrain. *Educational Researcher*, *33*(8), 3–15. doi:10.3102/0013189X033008003

Bowe, A. G., Desjardins, C. D., Covington Clarkson, L. M., & Lawrenz, F. (2015). Urban elementary single-sex math classrooms: Mitigating stereotype threat for African American girls. *Urban Education*, *52*(3), 370–398. doi:10.1177/0042085915574521

Bowman, J. D., Rogers, A. E. E., Monsalve, R. A., Mozdzen, T. J., & Mahesh, N. (2018). An absorption profile centred at 78 megahertz in the sky-averaged spectrum. *Nature*, *555*(7694), 67–70. doi:10.1038/nature25792 PMID:29493587

Boysen, G. (2012). Teacher and student perceptions of microaggressions in college classrooms. *College Teaching*, *60*(3), 122–129. doi:10.1080/87567555.2012.654831

Bragow, D., Gragow, K. A., & Smith, E. (1995). Back to the future: Toward curriculum integration. *Middle School Journal*, *27*(2), 39–46. doi:10.1080/00940771.1995.11496152

Bramwell-Lalor, S., & Rainford, M. (2014). The effects of using concept mapping for improving advanced level biology students' lower- and higher-order cognitive skills. *International Journal of Science Education*, *36*(5), 839–864. doi:10.1080/09500693.2013.829255

Bray, J. N. (2000). *Collaborative inquiry in practice: Action, reflection, and meaning making*. Thousand Oaks, CA: Sage Publications, Inc.

Bray, J. N. (2002). Uniting teacher learning: Collaborative inquiry for professional development. *New Directions for Adult and Continuing Education*, *2002*(94), 83–92. doi:10.1002/ace.62

Britner, S. L. (2008). Motivation in high school science students: A comparison of gender differences in life, physical, and earth science classes. *Journal of Research in Science Teaching*, *45*(8), 955–970. doi:10.1002/tea.20249

Brotman, J. S., & Moore, F. M. (2008). Girls and science. A review of four themes in the science education literature. *Journal of Research in Science Teaching*, *45*(9), 971–1002. doi:10.1002/tea.20241

Brown, A., & Green, T. (2016). Virtual reality: Low-cost tools and resources for the classroom. *TechTrends*, *60*(5), 517–519. doi:10.100711528-016-0102-z

Brown, B. A. (2004). Discursive identity: Assimilation into the culture of science and its implications for minority students. *Journal of Research in Science Teaching*, *41*(8), 810–834. doi:10.1002/tea.20228

Brown, B. A., Henderson, J. B., Gray, S., Donovan, B., & Sullivan, S. (2013). From Access to Success: Identity Contingencies & African-American Pathways to Science. *Higher Education Studies*, *3*(1). doi:10.5539/hes.v3n1p1

Brown, B. A., Henderson, J. B., Gray, S., Donovan, B., Sullivan, S., Patterson, A., & Waggstaff, W. (2016). From description to explanation: An empirical exploration of the African-American pipeline problem in STEM. *Journal of Research in Science Teaching*, *53*(1), 146–177. doi:10.1002/tea.21249

Bruning, M., Bystydzienski, J., & Eisenhart, M. (2012). Intersectionality as a framework for understanding diverse young women's interest in engineering. *2012 WEPAN National Conference*, 1-14.

Buck, G. A., Plano Clark, V. L., Leslie-Pelecky, D., Lu, Y., & Cerda-Lizarraga, P. (2008). Examining the cognitive processes used by adolescent girls and women in identifying science role models: A feminist approach. *Science Education*, *92*(4), 688–707. doi:10.1002ce.20257

Bulanda, J. R. (2011). Doing family, doing gender, doing religion: Structured ambivalence and the religion-family connection. *Journal of Family Theory & Review*, *3*(3), 179–197. doi:10.1111/j.1756-2589.2011.00093.x

Buldu, M. (2006). Young children's perceptions of scientists: A preliminary study. *Educational Research*, *48*(1), 121–132. doi:10.1080/00131880500498602

Bullock, E. C. (2017). *Only STEM Can Save Us? Examining Race, Place, and STEM Education as Property.* Retrieved from http://tandfonline.com/action/journalInformation?journalCode=heds20

Bullock, E. C. (2017). Only STEM can save us? Examining race, place, and STEM education as property. *Educational Studies*, *53*(6), 628–641. doi:10.1080/00131946.2017.1369082

Bunch, G. C. (2013). Pedagogical language knowledge: Preparing mainstream teachers for English learners in the new standards era. *Review of Research in Education*, *37*(1), 298–341. doi:10.3102/0091732X12461772

Bureau of Labor Statistics. (2010). *Occupational outlook handbook, 2010-11 Edition, agricultural and food scientists*. Retrieved from http://www.bls.gov/oco/ocos046.htm

Buschor, C. B., Berweger, S., Frei, A. K., & Kappler, C. (2014). Majoring in STEM—What accounts for women's career decision making? A mixed methods study. *The Journal of Educational Research*, *107*(3), 167–176. doi:10.1080/00220671.2013.788989

Buse, K., Bilimoria, D., & Perelli, S. (2013). Why they stay: Women persisting in US engineering careers. *Career Development International*, *18*(2), 139–154. doi:10.1108/CDI-11-2012-0108

Bybee, R. W., Taylor, J. A., Gardner, A., Van Scotter, P., Carlson Powell, J., Westbrook, A., & Landes, N. (2006). *BSCS 5E instructional model: Origins and effectiveness. A report prepared for the Office of Science Education, National Institutes of Health.* Colorado Springs, CO: BSCS.

Caleon, I. S., & Subramaniam, R. (2008). Attitudes towards science of intellectually gifted and mainstream upper primary students in Singapore. *Journal of Research in Science Teaching*, *45*(8), 940–954. doi:10.1002/tea.20250

Calvert, L. (2016). *Moving from compliance to agency: What teachers need to make professional learning work.* Oxford, OH: Learning Forward & NCTAF.

Camera, L. (2015). *Students in Urban School Districts Lagging in Math.* Reading, MA: U.S. News.

Capobianco, B. M., Diefes-Dux, H. A., Mena, I., & Weller, J. (2011). What is an engineer? Implications of elementary school student conceptions for engineering education. *Journal of Engineering Education*, *100*(2), 304–328. doi:10.1002/j.2168-9830.2011.tb00015.x

Capps, D. K., & Crawford, B. A. (2013). Inquiry-based instruction and teaching about nature of science: Are they happening? *Journal of Science Teacher Education*, *24*(3), 497–526. doi:10.100710972-012-9314-z

Capraro, R. M., Capraro, M. M., Scheurich, J. J., Jones, M., Morgan, J., Huggins, K. S., ... Han, S. (2016). Impact of sustained professional development in STEM on outcome measures in a diverse urban district. *The Journal of Educational Research*, *109*(2), 181–196. doi:10.1080/00220671.2014.936997

Carlone, H. B., & Johnson, A. (2007). Understanding the science experiences of successful women of color: Science identity as an analytic lens. *Journal of Research in Science Teaching*, *44*(8), 1187–1218. doi:10.1002/tea.20237

Carmichael, C. (2017, May 13). *A State-by-State Policy Analysis of STEM Education for K-12 Public Schools.* Seton Hall University Dissertaions and Theses.

Carnevale, A. P., Smith, N., & Melton, M. (2011). *STEM.* Washington, DC: Georgetown Universtiy Center on Education and the Workforce.

Carpi, A., Ronan, D. M., Falconer, H. M., & Lents, N. H. (2017). Cultivating minority scientists: Undergraduate research increases self-efficacy and career ambitions for underrepresented students in STEM. *Journal of Research in Science Teaching*, *54*(2), 169–194. doi:10.1002/tea.21341

Cavagnetto, A., Hand, B. M., & Norton-Meier, L. (2010). The nature of elementary student science discourse in the context of the science writing heuristic approach. *International Journal of Science Education*, *32*(4), 427–449. doi:10.1080/09500690802627277

Cavanagh, S. (2006). Simple Science Difficult for Urban Students to Grasp, NAEP Study Finds. *Education Week*.

CCSS Math - Common Core State Standards Initiative. (n.d.). Retrieved 5 29, 2018, from Common Core State Standards Initiative: http://www.corestandards.org/math

Ceci, S. J., & Williams, W. M. (2010). Sex differences in math-intensive fields. *Current Directions in Psychological Science*, *19*(5), 275–279. doi:10.1177/0963721410383241 PMID:21152367

Ceci, S. J., Williams, W. M., & Barnett, S. M. (2009). Women's underrepresentation in science: Sociocultural and biological considerations. *Psychological Bulletin*, *135*(2), 218–261. doi:10.1037/a0014412 PMID:19254079

Center for Applied Linguistics. (forthcoming). *Standards of professional development excellence for teachers of English learners.* Washington, DC: Author.

Chambers, D. (1983). Stereotypic images of the scientist: The draw-a-scientist test. *Science Education*, *67*(2), 255–265. doi:10.1002ce.3730670213

Chen, X. (2013). STEM attrition: College students' paths into and out of STEM fields. Washington, DC: Statistical Analysis Report NCES 2014-001.

Chen, H., Wang, H., Lin, H., Lawrenz, F. P., & Hong, Z. (2014). Longitudinal study of after-school, inquiry-based science intervention on low-achieving students and their perceptions of learning science. *International Journal of Science Education*, *13*(36), 2133–2156. doi:10.1080/09500693.2014.910630

Chen, J., & Cowie, B. (2014). Scientists talking to students through videos. *International Journal of Science and Mathematics Education*, *12*(2), 445–465. doi:10.100710763-013-9415-y

Cheryan, S., Master, A., & Meltzoff, A. N. (2015). Cultural stereotypes as gatekeepers: Increasing girls' interest in computer science and engineering by diversifying stereotypes. *Frontiers in Psychology*, *6*, 49. doi:10.3389/fpsyg.2015.00049 PMID:25717308

Cheung, D. (2007). Students' attitudes toward chemistry lessons: The interaction effect between grade level and gender. *Research in Science Education*, *39*(1), 75–91. doi:10.100711165-007-9075-4

Cho, F. (2012). The limits of the Buddhist embrace of science. *Science and Engineering Ethics*, *18*(3), 539–542. doi:10.100711948-012-9361-z PMID:22426623

Christidou, V. (2011). Interest, attitudes and images related to science: Combining students' voices with the voices of school science, teachers, and popular science. *International Journal of Environmental and Science Education*, *6*(2), 141–159.

Coalition for Science After School. (2007, March). *Blueprint for Action*. Retrieved from https://www.howtosmile.org/pdf/coalition%20publications/%20Science%20in%20After-School%20blueprint.pdf

Cobern, W., & Aikenhead, G. (1997). *Scientific Literacy and Cultural Studies Project*. Scientific Literacy and Cultural Studies Project. Paper 13.

Cochran-Smith, M., & Lytle, S. (1999). Relationships of knowledge and practice: Teacher learning in communities. *Review of Research in Education*, *24*, 249–305.

College and Career Readiness and Success Center. (2018). Retrieved from https://ccrscenter.org/ccrs-landscape/state-profile/district-columbia

Collier, K. L., Bos, H. M., Merry, M. S., & Sandfort, T. G. (2013). Gender, ethnicity, religiosity, and same-sex sexual attraction and the acceptance of same-sex sexuality and gender non-conformity. *Sex Roles*, *68*(11-12), 724–737. doi:10.100711199-012-0135-5 PMID:23687403

Collins, P. H. (2000). It's all in the family: Intersections of gender, race, and nation. In U. Narayan & S. Harding (Eds.), *Decentering the Center: Philosophy for a Multicultural, Postcolonial, and Feminist World* (pp. 156–176). Bloomington, IN: Indiana University Press.

Colvin, W., Lyden, S., & León de la Barra, B. A. (2012). Attracting girls to civil engineering through hands-on activities that reveal the communal goals and values of the profession. *Leadership and Management in Engineering*, *13*(1), 35–41. doi:10.1061/(ASCE)LM.1943-5630.0000208

Committee, U. S.-S. (2016). *Connecting the Dots: Sustaining a Culture of Assessment*. Washington, DC: UDC Self-Study Report, Prepared for the Middle States Commission On Higher Education and the UDC community.

Concannon, J. P., & Barrow, L. H. (2010). Men's and women's intentions to persist in undergraduate engineering degree programs. *Journal of Science Education and Technology, 19*(2), 133–145. doi:10.100710956-009-9187-x

Confrey, J., & Krupa, E. E. (2012). The Common Core State Standards for Mathematics: How Did We Get Here, and What Needs to Happen Next? In C. R. Hirsch, G. T. Lappan, & B. J. Reys (Eds.), Curriculum Issue in an Era of Common Core State Standards for Mathematics (pp. 3-15). The National Council of Teachers of Mathematics.

Confrey, J., & Krupa, E. E. (2010). *Curriculum Design, Development, and Implementation in an Era of Common Core State Standards: Summary Report of a Conferent.* Arlington, VA: Center for the Study of Mathematics Curriculum.

Cotabish, A., Dailey, D., Robinson, A., & Hughes, G. (2013). The effects of a STEM intervention on elementary students' science knowledge and skills. *School Science and Mathematics, 113*(5), 215–226. doi:10.1111sm.12023

Council of Chief State School Officers. (2012). *Framework for English language proficiency development standards corresponding to the Common Core State Standards and the Next Generation Science Standards.* Washington, DC: Author.

Counts, J., Katsiyannis, A., & Whitford, D. K. (2018). Culturally and linguistically diverse learners in special education: English learners. *NASSP Bulletin, 102*(1), 5–21. doi:10.1177/0192636518755945

Crenshaw, K. (1989). Demarginalizing the intersection of race and sex: A black feminist critique of antidiscrimination doctrine, feminist theory, and antiracist politics. *University of Chicago Legal Forum, 1989*, 139–167.

Crenshaw, K. (1991). Mapping the margins: Intersectionality, identity politics, and violence against women of color. *Stanford Law Review, 43*(6), 1241–1279. doi:10.2307/1229039

Cummins, J. (2000). *Language, power, and pedagogy: Bilingual children in the crossfire.* Clevedon, UK: Multilingual Matters. doi:10.21832/9781853596773

Cvencek, D., Meltzoff, A. N., & Greenwald, A. G. (2011). Math-gender stereotypes in elementary school children. *Child Development, 82*(3), 766–779. doi:10.1111/j.1467-8624.2010.01529.x PMID:21410915

Danter, E. H. (2005). *The intention-behavior gap: To what degree does Fishbein's integrated model of behavioral prediction predict whether teacher's implement material learned in a professional development workshop?* (Ed.D. Dissertation). Ohio State University. Retrieved from http://etd.ohiolink.edu/view.cgi?acc_num=osu1111698037

Darling-Hammond, L. (2000). Teacher Quality and Student Achievement. *Education Policy Analysis Archives*, *8*(1), 44.

Darling-Hammond, L., & Bransford, J. (Eds.). (2007). *Preparing teachers for a changing world: What teachers should learn and be able to do*. John Wiley & Sons.

Darling-Hammond, L., Hyler, M., & Gardner, M. (2017). *Effective Teacher Professional Development*. Palo Alto, CA: Learning Policy Institute.

Darling-Hammond, L., & McLaughlin, M. W. (1995). Policies that support professional development in an era of reform. *Phi Delta Kappan*, *76*(8), 597–604.

DC State Board of Education. (2018). *Teacher and principal turnover in public*. Retrieved from https://dcps.aqdev.dc.gov/page/compensation-lift-and-impactplus

DeArmond, M., Denice, P., Gross, B., Hernandez, J., Jochim, A., & Lake, R. (2015). *Measuring Up: Educational Improvement and Opportunity in 50 Cities*. Center on Reinventing Public Education.

Dee, T., & Goldhaber, D. (2017). *Understanding and Addressing*. The Hamilton Project.

Delandshere, G., & Petrosky, A. (2004). Political rationales and idelogical stances of the standards-based reform of teacher education in the U.S. *Teaching and Teacher Education*, *20*(1), 1–15. doi:10.1016/j.tate.2003.09.002

DePaul, A. (2000). *Survival Guide for New Teachers How New Teachers Can Work Effectively with Veteran Teachers, Parents, Principals, and Teacher Educators*. U.S. Department of Education, Office of Educational Research and Improvement.

Desimone, L. M., & Garet, M. S. (2015). Best practices in teachers' professional development in the United States. *Psychology. Social Education*, *7*(3), 252–263.

Desimone, L. M., Porter, A. C., Garet, M. S., Yoon, K. S., & Birman, B. F. (2002). Effects of professional development on teachers' instruction: Results from a three-year longitudinal study. *Educational Evaluation and Policy Analysis*, *24*(2), 81–112. doi:10.3102/01623737024002081

DeWitt, J., Archer, L., & Osborne, J. (2012). Nerdy, Brainy and Normal: Children's and Parents' Constructions of Those Who Are Highly Engaged with Science. *Research in Science Education*, *43*(4), 1455–1476. doi:10.100711165-012-9315-0

Dewitt, J., Osborne, J., Archer, L., Dillon, J., Willis, B., & Wong, B. (2011). Young children's aspirations in science: The unequivocal, the uncertain and the unthinkable. *International Journal of Science Education*, *33*, 1–50.

Dickson, P. (2001). *Sputnik: The shock of the century*. New York, NY: Walker Publishing.

DiDonato, M., Johnson, A., & Reisslein, M. (2014). A gender-specific, brochure-based intervention for improving boys' and girls' engineering stereotypes and academic self-perceptions. *Global Journal of Engineering Education*, *16*(1), 34–42.

Diekman, A. B., Brown, E. R., Johnston, A. M., & Clark, E. K. (2010). Seeking congruity between goals and roles a new look at why women opt out of science, technology, engineering, and mathematics careers. *Psychological Science*, *21*(8), 1051–1057. doi:10.1177/0956797610377342 PMID:20631322

Dionne, L., Reis, G., Trudel, L., Guillet, G., Kleine, L., & Hancianu, C. (2011). Students' sources of motivation for participating in science fairs: An exploratory study within the Canada-Wide Science Fair 2008. *International Journal of Science and Mathematics Education*, *10*(3), 669–693. doi:10.100710763-011-9318-8

Djonko-Moore, C. M., Leonard, J., Holifield, Q., Bailey, E. B., & Almughyirah, S. M. (2018). Using culturally relevant experiential education to enhance urban children's knowledge and engagement in science. *Journal of Experiential Education*, *41*(2), 137–153. doi:10.1177/1053825917742164

Doan, A. E., & Haider-Markel, D. P. (2010). The role of intersectional stereotypes on evaluations of gay and lesbian political candidates. *Politics & Gender*, *6*(1), 63–91. doi:10.1017/S1743923X09990511

Dortch, D., & Patel, C. (2017). Black undergraduate women and their sense of belonging in STEM at predominantly white institutions. *NASPA Journal About Women in Higher Education*, *10*(2), 202–215. doi:10.1080/19407882.2017.1331854

Dossey, J. A., McCrone, S. S., & Halvorsen, K. T. (2016). *Mathematics Education in the United States, 2016: A Capsule Summary Fact Book*. Reston, VA: National Council of Teachers of Mathematics.

Douglas, K. A., & Strobel, J. (2015). Hopes and goals survey for use in STEM elementary education. *International Journal of Technology and Design Education*, *25*(2), 245–259. doi:10.100710798-014-9277-9

DuFour, R., & Eaker, R. (1998). *Professional learning communities at work: Best practices for enhancing student achievement*. Bloomington, IN: National Educational Service.

Eagly, A. H., & Chaiken, S. (1993). *The psychology of attitudes*. London: Harcourt Brace Jovanovich College Publishers.

Earwood, E., & Suiter, P. (2012). *A scent of water: Bringing life back to the Christian school movement*. Greenville, SC: Ambassador International.

Eccles, J. (2009). Who am I and what am I going to do with my life? Personal and collective identities as motivators of action. *Educational Psychologist*, *44*(2), 78–89. doi:10.1080/00461520902832368

Eccles, J. (2011). Gendered educational and occupational choices: Applying the Eccles et al. model of achievement-related choices. *International Journal of Behavioral Development*, *35*(3), 195–201. doi:10.1177/0165025411398185

Eccles, J. S. (2007). Where are all the women? Gender differences in participation in physical science and engineering. In S. J. Ceci & W. M. Williams (Eds.), *Why aren't more women in science? Top researchers debate the evidence* (pp. 199–210). Washington, DC: American Psychological Association. doi:10.1037/11546-016

Eccles, J. S., & Jacobs, J. E. (1986). Social forces shape math attitudes and performance. *Signs (Chicago, Ill.)*, *11*(2), 367–380. doi:10.1086/494229

Echevarría, J., Richards-Tutor, C., Chinn, V. P., & Ratleff, P. A. (2011). Did they get it? The role of fidelity in teaching English learners. *Journal of Adolescent & Adult Literacy*, *54*(6), 425–434. doi:10.1598/JAAL.54.6.4

Echevarría, J., Vogt, M. E., & Short, D. (2017). *Making content comprehensible to secondary English learners: The SIOP model*. Boston: Allyn & Bacon.

Economics, F. (2018, November). *Economic Forecasts from the World's Leading Economists*. Retrieved from https://www.focus-economics.com/countries/united-states

Editorial, N. P. (2014). The Power of Analogies. *Nature Photonics*, *8*(1), 1. doi:10.1038/nphoton.2013.359

Education Commission of the States. (2014). *50-state comparison: What ELL training, if any, is required of general classroom teachers?* Available at http://ecs.force.com/mbdata/mbquestNB2?rep=ELL1415

Education Week Research Center. (2016). Quality Counts 2017: Under Construction—Building on ESSA's K-12 Foundation. *Education Week.*

Education Week. (2018, May 22). *Chart: Which STEM Jobs Are In Demand and Pay Well?* Retrieved from https://www.edweek.org/ew/section/multimedia/chart-which-stem-jobs-will-be-in.html

EducationD. O. (2011, June 6). Retrieved from https://osse.dc.gov/publication/math-common-core-crosswalk

Eisenhart, M., Weis, L., Allen, C. D., Cipollone, K., Stich, A., & Dominguez, R. (2015). High school opportunities for STEM: Comparing inclusive STEM-focused and comprehensive high schools in two US cities. *Journal of Research in Science Teaching*, *52*(6), 763–789. doi:10.1002/tea.21213

Ejiwale, J. (2003). Barriers To Successful Implementation of STEM Education. *Journal of Education and Learning*, *7*(2), 63–74.

Ellis, A., & Fouts, J. (2001). Interdisciplinary curriculum: The research base: The decision to approach music curriculum from an interdisciplinary perspective should include a consideration of all the possible benefits and drawbacks. *Music Educators Journal*, *87*(22), 22–26, 68. doi:10.2307/3399704

Else-Quest, N. M., Hyde, J. S., & Linn, M. C. (2010). Cross-national patterns of gender differences in mathematics: A meta-analysis. *Psychological Bulletin*, *136*(1), 103–127. doi:10.1037/a0018053 PMID:20063928

Ertmer, P. A., Richardson, J., Cramer, J., Hanson, L., Huang, W., Lee, Y., ... Um, E. J. (2005). Professional development coaches: Perceptions of critical characteristics. *Journal of School Leadership*, *15*(1), 52–75. doi:10.1177/105268460501500103

Etzioni, A., & Nunn, C. (1974). The public appreciation of science in contemporary America. *Daedalus*, *103*(3), 191–205.

Fairclough, N. (1992). *Language and power*. London: Longman.

Farland-Smith, D. (2012). Development and field test of the modified Draw-a-Scientist Test and the Draw-a-Scientist rubric. *School Science and Mathematics*, *112*(2), 109–116. doi:10.1111/j.1949-8594.2011.00124.x

Fellows, N. J. (1994). A window into thinking: Using student writing to understand conceptual change in science learning. *Journal of Research in Science Teaching*, *31*(9), 985–1001. doi:10.1002/tea.3660310911

Ferriman, K., Lubinski, D., & Benbow, C. P. (2009). Work preferences, life values, and personal views of topmath/science graduate students and the profoundly gifted: Developmental changes and gender differences during emerging adulthood and parenthood. *Journal of Personality and Social Psychology*, *97*(3), 517–532. doi:10.1037/a0016030 PMID:19686005

Finson, K., Pedersen, J., & Thomas, J. (2006). Comparing science teaching styles to students' perceptions of scientists. *School Science and Mathematics*, *106*(1), 8–15. doi:10.1111/j.1949-8594.2006.tb18066.x

Fishbein, M., & Ajzen, I. (1975). *Belief, attitude, intention and behavior: An introduction to theory and research*. Reading, MA: Addison-Wesley Publishing Co.

Fitzgerald, S. T., & Glass, J. L. (2014). *Religion and inequality in America: Research and theory on religion's role in stratification*. New York, NY: Cambridge University Press.

Flash, K., Allen, M., Mack, T., & Clement, K. (2017). STEM bridges: Evolution of an academic library STEM outreach program. *Journal of Library Administration*, *57*(8), 879–890. doi:10.1080/01930826.2017.1374105

Forbes, C. E., & Schmader, T. (2010). Retraining attitudes and stereotypes to affect motivation and cognitive capacity under stereotype threat. *Journal of Personality and Social Psychology*, *99*(5), 740–754. doi:10.1037/a0020971 PMID:20822288

Forman, J., Gubbins, E. J., Villanueva, M., Massicotte, C., Callahan, C., & Tofel-Grehl, C. (2015). National survey of STEM high schools' curricular and instructional strategies and practices. *NCSSS Journal*, *20*(1), 8–19.

Foster, G. M. (2013). Faith and race in American political life. *The Journal of American History*, *99*(4), 1206–1207. doi:10.1093/jahist/jas594

Freel, A. (1998). Achievement in urban schools: What makes the difference? *Education Digest*, *64*(1), 17.

Frenzel, M. W. (2008). *Reducing gender stereotypes in mathematics* (Master's Thesis). Retrieved from http://digitalcommons.brockport.edu/ehd_theses/234

Frome, P. M., Alfeld, C., Eccles, J., & Barber, B. L. (2006). Why don't they want a male-dominated job? An investigation of young women who changed their occupational aspirations. *Educational Research and Evaluation, 12*(4), 359–372. doi:10.1080/13803610600765786

Fullan, M. (2011). *Choosing the Wrong Drivers for Whole System Reform. Seminar Series Paper No. 204*. Victoria, Australia: Centre for Strategic Education.

Furner, J., & Kumar, D. (2007). The mathematics and science integration argument: A stand for teacher education. *Eurasia Journal of Mathematics. Science & Technology, 3*(3), 185–189.

Galguera, T. (2011). Participant structures as professional learning tasks and the development of pedagogical language knowledge among preservice teachers. *Teacher Education Quarterly, 38*, 85–106.

Gándara, P., Maxwell-Jolly, J., & Driscoll, A. (2005). *Listening to teachers of English language learners: A survey of California teachers' challenges, experiences, and professional development needs*. Santa Cruz, CA: The Center for the Future of Teaching and Learning.

García, E., & Weiss, E. (2017). *Education Inequalities at the School Starting Gate: Gaps, Trends, and Strategies to Address Them*. Economic Policy Institute.

García, S. B., Ortiz, A. A., & Sorrells, A. M. (2012). Intersectionality as a framework for research and practice in special education. *Multiple Voices for Ethnically Diverse Exceptional Learners, 13*(1), 1–3.

Garet, M. S., Porter, A. C., Desimone, L., Birman, B. F., & Yoon, K. S. (2001). What makes professional development effective? results from a national sample or teachers. *American Educational Research Journal, 38*(4), 915–945. doi:10.3102/00028312038004915

Gay, G. (2010). *Culturally responsive teaching: Theory, research, and practice* (2nd ed.). New York: Teachers College Press.

George, R. (2006). A cross-domain analysis of change in students' attitudes toward science and attitudes about the utility of science. *International Journal of Science Education, 28*(6), 571–589. doi:10.1080/09500690500338755

Gerber, B. L., Cavallo, M. L., & Edmund, A. M. (2001). Relationships among informal learning environments, teaching procedures and scientific reasoning ability. *International Journal of Sciences*, *23*(5), 535–549.

Giddens, A. (1991). *Modernity and self-identity: Self and society in the late modern age*. Stanford, CA: Stanford University Press.

Gist, C. D., Bianco, M., & Lynn, M. (2018). Examining Grow Your Own Programs Across the Teacher Development Continuum: Mining Research on Teachers of Color and Nontraditional Educator Pipelines. *Journal of Teacher Education*. doi:10.1177/0022487118787504

Glaser-Zikuda, M., & Fusz, S. (2008). Impact of teacher competencies on student emotions: A multi-method approach. *International Journal of Education and Research*, *47*(2), 136–147. doi:10.1016/j.ijer.2007.11.013

Glasman, L. R., & Albarracín, D. (2006). Forming attitudes that predict future behavior: A meta-analysis of the attitude-behavior relation. *Psychological Bulletin*, *132*(5), 778–822. doi:10.1037/0033-2909.132.5.778 PMID:16910754

Glen, N., & Dotger, S. (2013). Writing like a scientist: Exploring elementary teachers' understandings and practices of writing in science. *Journal of Science Teacher Education*, *24*(6), 957–976. doi:10.100710972-013-9348-x

Goldbrunner, T., Doz, Y., Wilson, K., & Veldhoen, S. (2006). The well-designed global R&D network. *Strategy and Business*, *43*, 15.

Goldman, A. D., & Penner, A. M. (2014). Exploring international gender differences in mathematics self-concept. *International Journal of Adolescence and Youth*, 1–16. PMID:27840545

Gonsalves, A., Rahm, J., & Carvalho, A. (2013). 'We could think of things that could be science': Girls' re-figuring of science in an out-of-school-time club. *Journal of Research in Science Teaching*, *50*(9), 1068–1097. doi:10.1002/tea.21105

Gordon, E. W. (2001). *Affirmative Development of Academic Abilities. Pedagogical Inquiry and Praxis. Number 2*. New York, NY: Columbia University.

Gottfredson, L. S. (2004). Applying Gottfredson's theory of circumspection and compromise in career guidance and counseling. In D. Brown & Lent (Eds.), Career Development and Counseling. San Francisco: John Wiley & Sons, Inc.

Gottfredson, L. (2006). Circumscription and compromise. In J. H. Greenhaus & G. A. Callanan (Eds.), *Encyclopedia of career development* (pp. 167–169). Thousand Oaks, CA: Sage Publications. doi:10.4135/9781412952675.n62

Gottfredson, L. S. (1981). Circumscription and compromise: A developmental theory of occupational aspirations. *Journal of Counseling Psychology*, *28*(6), 545–579. doi:10.1037/0022-0167.28.6.545

Gottfredson, L. S. (2002). Gottfredson's theory of circumscription, compromise, and self-creation. In D. Brown (Ed.), *Career Choice and Development* (4th ed.; pp. 85–148). San Francisco: Jossey-Bass.

Gottfredson, L. S. (2005). Applying Gottfredson's theory of circumscription and compromise in career guidance and counseling. In S. D. Brown & R. W. Lent (Eds.), *Career development and counseling: Putting theory and research to work* (pp. 71–100). Hoboken, NJ: John Wiley & Sons.

Gottfredson, L. S., & Lapan, R. (1997). Assessing gender-based circumscription of occupational aspirations. *Journal of Career Assessment*, *5*(4), 419–441. doi:10.1177/106907279700500404

Goulden, M., Frasch, K., & Mason, M. A. (2009). *Staying competitive: Patching America's leaky pipeline in the sciences*. Center for American Progress and the Berkeley Center on Health, Economic, & Family Security. Retrieved November 20, 2013 from www.americanprogress.org/issues/2009/11/pdf/women_and_sciences.pdf

Graham, K. J. (2018). *Catalyzing Change in High School Mathematics: Initiating Critical Conversations*. Reston, VA: The National Council of Teachers of Mathematics.

Granovskiy, B. (2018). *Science, technology, engineering, and mathematics (STEM) education: An overview*. Washington, DC: Congressional Research Service.

Grant, A. (2013). The Efficacy of Coaching. In J. Passmore, D. B. Peterson, & T. Freire (Eds.), *The Wiley-Blackwell Handbook of the Psychology of Coaching and Mentoring* (pp. 15–39). Hoboken, NJ: John Wiley & Sons.

Grant, C. A., & Zwier, E. (Eds.). (2014). *Intersectionality and urban education: Identities, policies, spaces & power*. Charlotte, NC: Information Age Publishers.

Greenberg, J., Walsh, K., & McKee, A. (2014). *NCTQ teacher prep review: A review of the nation's teacher preparation programs*. Washington, DC: National Council for Teacher Quality.

Green, L. S. (2014). Through the looking glass: Examining technology integration in school librarianship. *Knowledge Quest, 43*(1), 36–43.

Griffin, K. A. (2018, April 23). *Addressing STEM Culture and Climate to Increase Diversity in STEM Disciplines.* Retrieved from Higher Education Today: https://www.higheredtoday.org/2018/04/23/addressing-stem-culture-climate-increase-diversity-stem-disciplines/

Grossman, J. M., & Porche, M. V. (2013). Perceived gender and racial/ethnic barriers to STEM success. *Urban Education, 49*(6), 698–727. doi:10.1177/0042085913481364

Grossman, P., Wineburg, S., & Woolworth, S. (2001). Toward a theory of teacher community. *Teachers College Record, 103*(6), 942–1012. doi:10.1111/0161-4681.00140

Grover, S., & Pea, R. (2013). Computational thinking in K–12: A review of the state of the field. *Educational Researcher, 42*(1), 38–43. doi:10.3102/0013189X12463051

Guittar, N. A., & Pals, H. (2014). Intersecting gender with race and religiosity: Do unique social categories explain attitudes toward homosexuality? *Current Sociology, 62*(1), 41–62. doi:10.1177/0011392113516985

Gunderson, E., Ramirez, G., Levine, S., & Beilock, S. (2012). The role of parents and teachers in the development of gender-related math attitudes. *Sex Roles, 66*(3), 153–166. doi:10.100711199-011-9996-2

Gupta, M. (2004). Enhancing student performance through cooperative learning in physical sciences. *Assessment & Evaluation in Higher Education, 29*(1), 63–73. doi:10.1080/0260293032000158162

Gutherie, J. T., Wigfield, A., & VonSecker, C. (2000). Effects of integrated instruction on motivation and strategy use in reading. *Journal of Educational Psychology, 92*(2), 331–341. doi:10.1037/0022-0663.92.2.331

Halliday, M. A. K., & Martin, J. R. (1993). *Writing science: Literacy and discursive power.* Pittsburgh, PA: University of Pittsburgh Press.

Hand, B., Norton-Meier, L., Gunel, M., & Akkus, R. (2016). Aligning teaching to learning: A 3-year study examining the embedding of language and argumentation into elementary science classrooms. *International Journal of Science and Mathematics Education, 14*(5), 847–863. doi:10.100710763-015-9622-9

Hand, B., Shelley, M. C., Laugerman, M., Fostvedt, L., & Therrien, W. (2018). Improving critical thinking growth for disadvantaged groups within elementary school science: A randomized controlled trial using the science writing heuristic approach. *Science Education, 102*(4), 693–710. doi:10.1002ce.21341

Hanley, P., Bennett, J., & Ratcliffe, M. (2014). The inter-relationship of science and religion: A typology of engagement. *International Journal of Science Education, 36*(7), 1210–1229. doi:10.1080/09500693.2013.853897

Hannah, J. S., & Kahn, S. E. (1989). The relationship of socioeconomic status and gender to the occupational choices of Grade 12 students. *Journal of Vocational Behavior, 34*(2), 161–178. doi:10.1016/0001-8791(89)90012-2

Hansen, M., & Gonzalez, T. (2014). Investigating the Relationship between STEM Learning Principles and Student Achievement in Math and Science. *American Journal of Education, 120*(2), 139–171. doi:10.1086/674376

Hanson, S. L. (2012). Science for all? The intersection of gender, race and science. *The International Journal of Science in Society, 3*(2), 113–136. doi:10.18848/1836-6236/CGP/v03i02/51325

Hanson, S. L. (2013). STEM degrees and occupations among Latinos: An examination of race/ethic and gender variation. *Race, Gender, & Class, 20*(1-2), 214–231.

Haq, R. (2013). Intersectionality of gender and others forms of identity: Dilemmas and challenges facing women in India. *Gender in Management, 28*(3), 171–184. doi:10.1108/GM-01-2013-0010

Hardre, P. L., & Sullivan, D. W. (2008). Student differences and environment perceptions: How they contribute to student motivation in rural high schools. *Learning and Individual Differences, 18*(4), 471–485. doi:10.1016/j.lindif.2007.11.010

Harper, N. W., & Daane, C. J. (1998). Causes and reduction of math anxiety in preservice elementary. *Action in Teacher Education, 19*(4), 29–38. doi:10.1080/01626620.1998.10462889

Harrison, C., & Tanner, K. D. (2018). Language Matters: Considering Microaggressions in Science. *CBE Life Sciences Education, 17*(1), fe4. doi:10.1187/cbe.18-01-0011 PMID:29496676

Harwell, M., Moreno, M., Phillips, A., Guzey, S. S., Moore, T. J., & Roehrig, G. H. (2015). A study of STEM assessments in engineering, science, and mathematics for elementary and middle school students. *School Science and Mathematics*, *115*(2), 66–74. doi:10.1111sm.12105

Hawkins, J. D., Guo, J., Hill, K. G., Battin-Pearson, S., & Abbott, R. D. (2001). Long-term effects of the Seattle social development intervention on school bonding trajectories. *Applied Developmental Science*, *5*(4), 225–236. doi:10.1207/S1532480XADS0504_04 PMID:17955057

Haworth, C. M., Dale, P., & Plomin, R. (2008). A twin study into the genetic and environmental influences on academic performance in science in nine-year-old boys and girls. *International Journal of Science Education*, *30*(8), 1003–1025. doi:10.1080/09500690701324190 PMID:20111670

Haydon, T., & Musti-Rao, S. (2011). Effective use of behavior-specific praise: A middle school case study. *Beyond Behavior*, *20*(2), 31–39.

Heath, S. B. (1982). Questioning at home and at school: A comparative study. *Doing the ethnography of schooling: Educational anthropology in action,* 102-131.

Heath, S. B. (1983). *Ways with words: Language, life, and work in communities and classrooms*. Cambridge, UK: Cambridge University Press. doi:10.1017/CBO9780511841057

Hein, G. (2006). Science education for a thriving democracy. *Hands On! 29*(1), 4–7, 19.

Helms, J. E. (1992). *A race is a nice thing to have: A guide to being a white person or understanding the white persons in your life*. Topeka, KS: Content Communications.

Helwig, A. (1998). Gender-role stereotyping: Testing theory with a longitudinal sample. *Sex Roles*, *38*(5/6), 403–423. doi:10.1023/A:1018757821850

Henward, A., & MacGillivray, L. (2012). It's a combination of the Bible and what's in your heart: Unresolvable tensions and contested narratives in a southern child care center. *Journal of Curriculum Theorizing*, *28*(3), 91–108.

Hibpshman, T. (2004). *A review of value-added models*. Frankfort, KY: Kentucky Education Professional Standards Board.

Highfield, R. (2011). Science presenter: Why is the popular image of scientists stuck in the past? *New Scientist*, *211*(2829), 29–30. doi:10.1016/S0262-4079(11)62210-0

Hill, C., Corbett, C., & St. Rose, A. (2010). *Why so few? Women in science, technology, engineering, and mathematics*. Washington, DC: American Association of University Women.

Hill, H. (2009). Fixing teacher professional development. *Phi Delta Kappan, 90*(7), 470–476. doi:10.1177/003172170909000705

Hill, P. D., Friedland, E. S., & Phelps, S. (2012). How Teacher Candidates' Perceptions of Urabn Students Are Influenced by Field Experiences: A Review of the Literature. *Action in Teacher Education, 34*(1), 77–96. doi:10.1080/01626620.2012.642290

Hird, M. D. (2013). *Barriers to implementation of new programs and pedagogies in K-12 STEM Education: a systems perspective*. Retrieved from http://hdl.handle.net/1721.1/81121

Hirsch, L. S., Berliner-Heyman, S., Cano, R., Kimmel, H., & Carpinelli, J. (2011, October). *Middle school girls' perceptions of engineers before and after a female only summer enrichment program*. Paper presented at the Frontiers in Education Conference, Rapid City, SD. 10.1109/FIE.2011.6142990

Hmielowski, J. D., Feldman, L., Myers, T. A., Leiserowitz, A., & Maibach, E. (2013). An attack on science? Media use, trust in scientists, and perceptions of global warming. *Public Understanding of Science (Bristol, England), 22*(3), 1–18. PMID:23825287

Hoh, Y. K. (2009). Using notable women in environmental engineering to dispel misperceptions of engineers. *International Journal of Environmental and Science Education, 4*(2), 117–131.

Holcomb-McCoy, C. (2007). *School counseling to close the achievement gap: A social justice framework for success*. Thousand Oaks, CA: Corwin.

Holland, D., Lachicotte, W., Skinner, D., & Cain, C. (2001). *Identity and agency in cultural worlds*. Cambridge, MA: Harvard University Press.

Hollins, E., & Guzman, M. T. (2005). Research on preparing teachers for diverse populations. In M. Cochran-Smith & K. Zeichner (Eds.), *Studying teacher education: The report of the AERA panel on research and teacher education* (pp. 477–548). Mahwah, NJ: Lawrence Erlbaum.

Horizon Research, Inc. (2013). *2012 National survey of science and mathematics education: Highlights Report*. Chapel Hill, NC: Author.

Hoven, M. (2013). A critical guide to resources in religion and science. *Religious Education (Chicago, Ill.)*, *108*(3), 280–297. doi:10.1080/00344087.2013.783361

Howard, T. C., & Milner, H. R. (2014). Teacher preparation for urban schools. In R. Milner & K. Lomotey (Eds.), *Handbook of urban education* (pp. 199–216). New York, NY: Routledge.

Hudicourt-Barnes, J. (2003). The use of argumentation in Haitian Creole science classrooms. *Harvard Educational Review*, *73*(1), 73–93. doi:10.17763/haer.73.1.hnq801u57400l877

Hudley, C. (2013, October). Education and urban schools. *The SES Indicator, 6*(2).

Huffling, L. D., Benavides, A., Ash, M., Germuth, A., Matthews, C., & Carlone, H. B. (2017). Promoting equitable access to STEM in a summer herpetology fieldwork program. (science, technology, engineering and mathematics). *Science Teacher (Normal, Ill.)*, *84*(6), 62. Retrieved from https://search.proquest.com/docview/1933244150

Huguet, P., & Régner, I. (2007). Stereotype threat among schoolgirls in quasi-ordinary classroom circumstances. *Journal of Educational Psychology*, *99*(3), 545–560. doi:10.1037/0022-0663.99.3.545

Hurley, M. (2001). Reviewing integrated science and mathematics: The search for evidence and definitions from new perspectives. *School Science and Mathematics*, *101*, 259–268. doi: .2001.tb18028.x doi:10.1111/j.1949-8594

Hussar, W. J., & Bailey, T. M. (2016). *Projection of Education Statistics to 2024 (NCES2016-013)*. Washington, DC: U.S. Department of Education, National Center for Education Statistics.

Huynh, V. H. (2012). Ethnic microaggressions and the depressive and somatic symptoms of Latino and Asian American adolescents. *Journal of Youth and Adolescence*, *41*(7), 831–846. doi:10.100710964-012-9756-9 PMID:22453294

Icel, M. (2018). Implementation of STEM policy: A case study of a STEM-focused urban charter school. *Journal of STEM Education*, *19*(3), 7–13.

Indiogine, S. P. (2015). The Achievement Gaps in Mathematics and Science. In A. Sahin (Ed.), *A Practice-based Model of STEM Teaching* (pp. 15–28). Rotterdam: SensePublishers. doi:10.1007/978-94-6300-019-2_2

Initiative, C. C. (2010). *Common Core State StandardS*. Retrieved 5 29, 2018, from Common Core State Standards Initiative: http://www.corestandards.org/assets/CCSSI_Math%20Standards.pdf

Inoue, N. (2007). Why face a challenge? The reason behind intrinsically motivated students' spontaneous choice of challenging tasks. *Learning and Individual Differences*, *17*(3), 251–259. doi:10.1016/j.lindif.2007.02.002

Institute for the Promotion of Teaching Science and Technology (IPST). (2008). *National science curriculum standards: The basic education core curriculum B.E.25541*. Retrieved November 13, 2013, from ww.ipst.ac.th/web/images/files/SciCur_2008_EngVersion.pdf

Isik-Ercan, Z., Zeynep Inan, H., Nowak, J. A., & Kim, B. (2014). 'We put on the glasses and moon comes closer!' urban second graders exploring the earth, the sun and moon through 3D technologies in a science and literacy unit. *International Journal of Science Education*, *36*(1), 129–156. doi:10.1080/09500693.2012.739718

Iskander, M., & Kapila, V. (2011). Revitalizing achievement by using instrumentation in science education (RAISE), a GK-12 fellows project. *Journal of Professional Issues in Engineering Education and Practice*, *138*(1), 62–72. doi:10.1061/(ASCE)EI.1943-5541.0000085

Johnson, A. M., Ozogul, G., DiDonato, M. D., & Reisslein, M. (2013). Engineering perceptions of female and male K-12 students: Effects of a multimedia overview on elementary, middle-, and high-school students. *European Journal of Engineering Education*, *38*(5), 519–531. doi:10.1080/03043797.2013.811477

Johnson, C. C., Zhang, D., & Kahle, J. B. (2012). Effective science instruction: Impact on high-stakes assessment performance. *RMLE Online*, *35*(9), 1–14. doi:10.1080/19404476.2012.11462092

Joyce, B. R., & Showers, B. (2002). *Designing Training and Peer Coaching: Our needs for learning*. ASCD.

Kagan, S., Kagan, M., & Kagan, L. (2016). *59 Kagan Structures: Proven Engagement Strategies*. San Clemente, CA: Kagan.

Kanny, M. A., Sax, L. J., & Riggers-Piehl, T. A. (2014). Investigating forty years of STEM research: How explanations for the gender gap have evolved over time. *Journal of Women and Minorities in Science and Engineering*, *20*(2), 127–148. doi:10.1615/JWomenMinorScienEng.2014007246

Kanwar, R. (2010). Sustainable water systems for agriculture and 21st century challenges. *Journal of Crop Improvement*, *24*(1), 41–59. doi:10.1080/15427520903307700

Karatas, F. O., Micklos, A., & Bodner, G. M. (2011). Sixth-grade students' views of the nature of engineering and images of engineers. *Journal of Science Education and Technology*, *20*(2), 123–135. doi:10.100710956-010-9239-2

Katz, I., & Assor, A. (2007). When choice motivates and when it does not. *Educational Psychology Review*, *19*(4), 429–442. doi:10.100710648-006-9027-y

Keeley, P. (2009). Elementary science education in the K-12 System. *NSTA WebNews Digest*. Retrieved August 23, 2013 from http://www.nsta.org/publications/news/story.aspx?id=55954

Kerr, B., Multon, K., Syme, M., Fry, N., Owens, R., Hammond, M., & Robinson-Kurpius, S. (2012). Development of the distance from privilege measures: A tool for understanding the persistence of talented women in STEM. *Journal of Psychoeducational Assessment*, *30*(1), 88–102. doi:10.1177/0734282911428198

Kessels, U., Heyder, A., Latsch, M., & Hannover, B. (2014). How gender differences in academic engagement relate to students' gender identity. *Educational Research*, *56*(2), 220–229. doi:10.1080/00131881.2014.898916

Kimmel, L., Miller, J., & Eccles, J. (2012). Do the paths to STEMM professions differ by gender? *Peabody Journal of Education*, *87*(1), 92–113. doi:10.1080/0161956X.2012.642276

Kind, P., Jones, K., & Barmby, P. (2007). Developing attitudes towards science measures. *International Journal of Science Education*, *29*(7), 871–893. doi:10.1080/09500690600909091

King, K., & Wiseman, D. (2001). Comparing science efficacy beliefs of elementary education majors in integrated and non-integrated teacher education coursework. *Journal of Science Teacher Education*, *12*(2), 143–153. doi:10.1023/A:1016681823643

Knight, M., & Cunningham, C. (2004, June). *Draw an engineer test (DAET): Development of a tool to investigate students' ideas about engineers and engineering*. Paper presented at 2004 American Society for Engineering Education Annual Conference & Exposition, Salt Lake City, UT.

Knight, M., & Cunningham, C. M. (2004). Draw an Engineer Test (DAET): Development of a tool to investigate students' ideas about engineers and engineering. American Society of Engineering Education.

Knudsen, S. (2006). Intersectionality-A theoretical inspiration in the analysis of minority cultures and identities in textbooks. In E. Bruillar, B. Aamotsbakken, S. Knudzen, & M. Horsley (Eds.), Caught in the Web or Lost in the Textbook (pp. 61–76). Academic Press. Retrieved from http://www.iartem.no/documents/caught_in_the_web.pdf#page=61

Koballa, T. R., & Glynn, S. M. (2007). Attitudinal and motivational constructs in science learning. In S. K. Abell & N. Lederman (Eds.), *Handbook for research in science education* (pp. 75–102). Mahwah, NJ: Lawrence Erlbaum Associates, Inc.

Koch, J. (2007). A gender inclusive approach to science education. In D. Sadker & E. Silber (Eds.), *Gender in the classroom: Foundations, skills, methods, and strategies across the curriculum* (pp. 205–223). Mahwah, NJ: Lawrence Erlbaum Associates.

Kohli, R., & Solórzano, D. G. (2012). Teachers, please learn our names!: Racial microagressions and the K-12 classroom. *Race, Ethnicity and Education*, *15*(4), 441–462. doi:10.1080/13613324.2012.674026

Koppal, M., & Caldwell, A. (2004). Meeting the Challenge of Science Literacy: Project 2061 Efforts to Improve Science Education. *Cell Biology Education*, *3*(1), 28–30. doi:10.1187/cbe.03-10-0016 PMID:22039343

Koul, R., Lerdpornkulrat, T., & Chantara, S. (2011). Relationship between career aspirations and measures of motivation toward biology and physics, and the influence of gender. *Journal of Science Education and Technology*, *20*(6), 761–770. doi:10.100710956-010-9269-9

Kowal, J., & Steiner, L. (2007). *Instructional Coaching. Issue Brief.* Center for Comprehensive School Reform and Improvement.

Kraft, M. A., Blazar, D., & Hogan, D. (2018). The effect of teacher coaching on instruction and achievement: A meta-analysis of the causal evidence. *Review of Educational Research*, *88*(4), 547–588. doi:10.3102/0034654318759268

Kramer, D. (2008). Sea perch is a fun, hands-on approach to teaching science. *Physics Today*, *61*(3), 22–23. doi:10.1063/1.2897940

Krishamurthi, A. (2014). *Examining the impact of afterschool STEM programs.* Academic Press.

Krogh, L., & Andersen, H. (2013). "Actually, I may be clever enough to do it". Using identity as a lens to investigate students' trajectories towards science and university. *Research in Science Education*, *43*(2), 711–731. doi:10.100711165-012-9285-2

Lacey, T., & Wright, B. (2009). Employment outlook: 2008-18 Occupational employment projections to 2018. *Monthly Labor Review*, *132*, 82.

Langdon, D., McKittrick, G., Beede, D., Khan, B., & Doms, M. (2011). *STEM: Good jobs now and for the future*. U.S. Department of Commerce, Economics and Statistics Administration. Retrieved from www.esa.doc.gov/sites/default/files/reports/documents/stemfinalyjuly14_1.pdf

Learning Forward. (2012). *Standards for professional learning*. Retrieved from: www.learningforward.org/standards/index.cfm

Lee, O., Maerten-Rivera, J., Penfield, R. D., LeRoy, K., & Secada, W. (2008). Science achievement of English language learners in urban elementary schools: Results of a first-year professional development intervention. *Journal of Research in Science Teaching*, *45*(1), 31–52. doi:10.1002/tea.20209

Lemke, J. L. (1990). *Talking Science: Language, Learning, and Values*. Norwood, NJ: Ablex Publishing Corporation.

Lent, R. W., Brown, S. D., & Hackett, G. (1994). Toward a unifying social cognitive theory of career and academic interest, choice, and performance. *Journal of Vocational Behavior*, *45*(1), 79–122. doi:10.1006/jvbe.1994.1027

Lester, J., Yamanaka, A., & Struthers, B. (2016). Gender microaggressions and learning environments: The role of physical space in teaching pedagogy and communication. *Community College Journal of Research and Practice*, *49*(11), 909–926. doi:10.1080/10668926.2015.1133333

Levine, M., Serio, N., Radaram, B., Chaudhuri, S., & Talbert, W. (2015). Addressing the STEM gender gap by designing and implementing an educational outreach chemistry camp for middle school girls. *Journal of Chemical Education*, *92*(10), 1639–1644. doi:10.1021/ed500945g

Levine-Rasky, C. (2011). Intersectionality theory applied to whiteness and middle-classness. *Social Identities*, *17*(2), 239–253. doi:10.1080/13504630.2011.558377

Lewis, L., Snow, K., Farris, E., Smerdon, B., Cronen, S., & Kaplan, J. (1999). Condition of America's public school facilities. *Education Statistics Quarterly*, *2*, 42–46.

Liben, L. S., & Coyle, E. F. (2014). Developmental interventions to address the STEM gender gap: Exploring intended and unintended consequences. *Advances in Child Development and Behavior*, *47*, 77–115. doi:10.1016/bs.acdb.2014.06.001 PMID:25344994

Lightbown, P. M., & Spada, N. (2006). *How languages are learned*. Oxford, UK: Oxford University Press.

Lindahl, K., & Watkins, N. M. (2015). Creating a culture of language awareness in content-based contexts. *TESOL Journal*, *6*(4), 777–789. doi:10.1002/tesj.223

Lindner, J. R., Rayfield, J., Briers, G., & Johnson, L. (2012). Graduate student fellowship program effects on attitude and interest toward science of middle school students. *Journal of Natural Resources and Life Sciences Education*, *41*(1), 15–21. doi:10.4195/jnrlse.2011.0021k

Lippman, L., Burns, S., & McArthur, E. (1996). *Urban Schools: The Challenge of Location and Poverty*. Nationa Center for Education Statistics.

Lloyd, S. D., & Harwin, A. (2016). Nation's Schools Get Middling Grade on Quality Counts Report Card. *Education Week*.

Logan, J. R., & Burdick-Will, J. (2017). School segregation and disparities in urban, suburban, and rural areas. *The Annals of the American Academy*, *674*(1), 199–216. doi:10.1177/0002716217733936 PMID:29430018

Loveless, T. (2013). *How well are American students learning? With sections on the latest international tests, tracking, and ability grouping, and advanced math 8th grade*. The Brown Center on Educational Policy.

Lucas, T. (2011). Language, schooling, and the preparation of teachers for linguistic diversity. In T. Lucas (Ed.), *Teacher preparation for linguistically diverse classrooms: A resource for teacher educators* (pp. 3–17). New York, NY: Routledge.

Maddock, M. N. (1981). Science education: An anthropological viewpoint. *Studies in Science Education*, *8*, 1–26.

Maltese, A. V., & Tai, R. H. (2010). Eyeballs in the fridge: Sources of early interest in science. *International Journal of Science Education*, *32*(5), 669–685. doi:10.1080/09500690902792385

Marzano, R. J., Simms, J. A., Roy, T., Heflebower, T., & Warrick, P. B. (2013). *Coaching classroom instruction*. Bloomington, IN: Marzano Research Laboratory.

Mason, C., Kahle, J., & Gardner, A. (1991). Draw-a-scientist test: Future implications. *School Science and Mathematics*, *91*(5), 193–198. doi:10.1111/j.1949-8594.1991.tb12078.x

Master, A., Cheryan, S., & Meltzoff, A. N. (2016). Computing whether she belongs: Stereotypes undermine girls' interest and sense of belonging in computer science. *Journal of Educational Psychology*, *108*(3), 424–437. doi:10.1037/edu0000061

Master, A., & Meltzoff, A. N. (2016). Building bridges between psychological science and education: Cultural stereotypes, STEM, and equity. *Prospects*, *46*(2), 215–234. doi:10.100711125-017-9391-z

Mathis, W. (2010). *Education and the Public Interest Center & Education Policy Research Unit*. Retrieved from http://epicpolicy.org/publication/common-core-standards

Matlen, B. J., Vosniadou, S., Jee, B., & Ptouchkina, M. (2011). Enhancing the comprehension of science text through visual analogies. In L. Carlson, C. Holscher, & T. Shipley (Eds.), *Proceedings of the 34th Annual Conference of the Cognitive Science Society* (pp. 2910–2915). Academic Press.

Matthew, R. (2017, May 10). *Students Shouldn't Live in STEM Deserts*. Retrieved from U.S. news and World Report: https://www.usnews.com/opinion/knowledge-bank/articles/2017-05-10/the-us-must-address-disparities-in-access-to-stem-education

Matthews, L. E. (2008, December). Illuminating Urban Excellence: A Movement of Change with Mathematics Education. *Journal of Urban Mathematics Education*, *11*(1-2), 7–10. http://education.gsu.edu/JUME Retrieved December 14, 2018

Ma, X., & Johnson, W. (2008). Mathematics as the critical filter: Curricular effects on gendered career choices. In H. M. G. Watt & J. S. Eccles (Eds.), *Gender and occupational outcomes: Longitudinal assessments of individual, social, and cultural influences* (pp. 55–83). Washington, DC: American Psychological Association. doi:10.1037/11706-002

Maxwell, D. O., Lambeth, D. T., & Cox, J. T. (2015). Effects of using inquiry-based learning on science achievement for fifth-grade students. *Asia-Pacific Forum on Science Learning & Teaching*, *1*(16), 106–136.

McAnn, A. (2018, July 30). *States with the Best & Worst School Systems*. Retrieved from www.Wallethub.com: https://wallethub.com/edu/states-with-the-best-schools/5335/

McGee, E. O. (2016). Devalued Black and Latino Racial Identities. *American Educational Research Journal*, *53*(6), 1626–1662. doi:10.3102/0002831216676572

McHenry, N., & Borger, L. (2013). How can teacher-directed professional development lead to the identification, utilization, reflection on, and revision of 5E Learning Progressions? *Electronic Journal of Science Education*, *17*(1), 1–24.

McKown, C., & Strambler, M. J. (2009). Developmental antecedents and social and academic consequences of stereotype-consciousness in middle childhood. *Child Development*, *80*(6), 1643–1659. doi:10.1111/j.1467-8624.2009.01359.x PMID:19930343

McKown, C., & Weinstein, R. S. (2003). The development and consequences of stereotype consciousness in middle childhood. *Child Development*, *74*(2), 498–515. doi:10.1111/1467-8624.7402012 PMID:12705569

Mead, M., & Metraux, R. (1957). The image of the scientist among high school students: A pilot study. *Science*, *126*(3270), 384–390. doi:10.1126cience.126.3270.384 PMID:17774477

Meguid, E. A., & Collins, M. (2017). Students' perceptions of lecturing approaches: traditional versus interactive teaching. *Advances in Medical Education and Practices*, 229-241.

Mercer, S. H., Zeigler-Hill, V., Wallace, M., & Hayes, D. M. (2011). Development and initial validation of the Inventory of Microaggressions Against Black Individuals. *Journal of Counseling Psychology*, *58*(4), 457–469. doi:10.1037/a0024937 PMID:21842986

Mervis, J. (2011, September 14). Is There a Special Formula for Successful STEM Schools? *Science Insider.* Retrieved from http://news.sciencemag.org/scienceinsider/2011/05/-is-there-a-special-formula-for-.html

MET Project Research Paper. (2010). *Learning About Teaching: Initial Findings from the Measures of Effective Teaching Project.* Bill and Melinda Gates Foundation.

Meyer, X., & Crawford, B. A. (2011). Teaching science as a cultural way of knowing: Merging authentic inquiry, nature of science, and multicultural strategies. *Cultural Studies of Science Education*, *6*(3), 525–547. doi:10.100711422-011-9318-6

Michaels, S. (2006). Narrative presentations: An oral preparation for literacy with first graders. In J. Cook-Gumperz (Ed.), *The Social Construction of Literacy* (2nd ed.; pp. 94–116). New York: Cambridge University Press. doi:10.1017/CBO9780511617454.006

Microsoft Corp. & Harris Interactive. (2011). *STEM Perceptions: Student & Parent Study Parents and Students Weigh in on How to Inspire the Next Generation of Doctors, Scientists, Software Developers and Engineers.* Harris Interactive.

Milam, J. (2012). *Girls and STEM education: A literature review*. Atlanta, GA: Georgia Institute of Technology.

Miller, J. D., & Kimmel, L. G. (2012). Pathways to a STEMM profession. *Peabody Journal of Education, 87*(1), 26–45. doi:10.1080/0161956X.2012.642274

Milner, H. R. IV. (2012, May 17). But What is Urban Education. *Urban Education, 47*(3), 556–561. doi:10.1177/0042085912447516

Minner, D. D., Levy, A. J., & Century, J. (2010). Inquiry-based science instruction—What is it and does it matter? Results from a research synthesis years 1984 to 2002. *Journal of Research in Science Teaching, 47*(4), 474–496. doi:10.1002/tea.20347

Minogue, J., Madden, L., Bedward, J., Wiebe, E., & Carter, M. (2010). The cross-case analyses of elementary students' engagement in the strands of science proficiency. *Journal of Science Teacher Education, 21*(5), 559–587. doi:10.100710972-010-9195-y

Mirza, H. S. (2013). A second skin: Embodied intersectionality, transnationalism, and narratives of identity and belonging among Muslim women in Britain. *Women's Studies International Forum, 36*, 5–15. doi:10.1016/j.wsif.2012.10.012

Mooney, C. (2006). *The Republican War on Science*. New York, NY: Basic Books.

Mooney, C. (2012). *The Republican Brain: The Science of Why They Deny Science—and Reality*. Hoboken, NJ: John Wiley & Sons.

Morgan, C., Isaac, J., & Sansone, C. (2001). The role of interest in understanding the career choices of female and male college students. *Sex Roles, 44*(5), 295-320. Retrieved from https://search.proquest.com/docview/225379950

Morgan, M., Farkas, G., Hillemeier, M. M., & Maczuga, S. (2016). Science Achievement Gaps Begin Very Early, Persist, and Are Largely Explained by Modifiable Factors. *Educational Researcher, 45*(1), 18–35. doi:10.3102/0013189X16633182

Morgan, S. L., Gelbgiser, D., & Weeden, K. A. (2013). Feeding the pipeline: Gender, occupational plans, and college major selection. *Social Science Research, 42*(4), 989–1005. doi:10.1016/j.ssresearch.2013.03.008 PMID:23721669

Morley, L. (2013). The rules of the game: Women and the leaderist turn in higher education. *Gender and Education*, *25*(1), 116–131. doi:10.1080/09540253.2012.740888

Mortensen, M. F., & Smart, K. (2007). Free-choice worksheets increase students' exposure to curriculum during museum visits. *Journal of Research in Science Teaching*, *44*(9), 1389–1414. doi:10.1002/tea.20206

Museus, S. D., Palmer, R. T., Davis, R. J., & Maramba, D. C. (2011). *Racial and ethnic minority students' success in STEM education. ASHE-Higher Education Report Series*. Hoboken, NJ: Jossey-Bass.

Museus, S., Palmer, R. T., Davis, R. J., & Maramba, D. C. (2011). Racial and ethnic minority students' success in STEM education. *ASHE Higher Education Report*, *36*(6), 1–136.

Nadal, K. L., Issa, M., Leon, J., Meterko, V., Wideman, M., & Wong, Y. (2011). Sexual orientation microaggressions: "death by a thousand cuts" for lesbian, gay, and bisexual youth. *Journal of LGBT Youth*, *8*(3), 234–259. doi:10.1080/19361653.2011.584204

Nadal, K. L., Sriken, J., Davidoff, K. C., Wong, Y., & McLean, K. (2013). Microaggressions within families: Experiences of multiracial people. *Family Relations*, *62*(1), 190–201. doi:10.1111/j.1741-3729.2012.00752.x

Nadal, K. L., Wong, Y., Griffin, K. E., Davidoff, K., & Sriken, J. (2014). The adverse impact of racial microaggressions on college students' self-esteem. *Journal of College Student Development*, *55*(5), 461–474. doi:10.1353/csd.2014.0051

Nadelson, L. S., Seifert, A., Moll, A. J., & Coats, B. (2012). i-STEM summer institute: An integrated approach to teacher professional development in STEM. *Journal of STEM Education: Innovation and Outreach*.

Najafi, M., Ebrahimitabass, E., Dehghani, A., & Rezaei, M. (2012). Students' attitude toward science and technology. *Interdisciplinary Journal of Contemporary Research in Business*, *3*(10), 129–134.

Nasir, N. S., & Vakil, S. (2017). STEM-focused academies in urban schools: Tensions and possibilities. *Journal of the Learning Sciences*, *26*(3), 376–406. doi:10.1080/10508406.2017.1314215

National Academy of Sciences. (2006). *Rising above the gathering storm: Energizing and employing America for a brighter economic future*. Washington, DC: National Academies Press.

National Academy of Sciences. (2010). *Rising Above the Gathering Storm, Revisited: Rapidly Approaching Category 5*. Washington, DC: The National Academies Press. Retrieved from http://www.uic.edu/home/ Chancellor/risingabove.pdf

National Academy of Sciences. (2011). *Joint G8+ science academies' statement on education for a science-based global development*. Washington, DC: The National Academies Press. Retrieved from http://www.nationalacademies.org/ includes/ Final_Education.pdf

National Assessment for Educational Progress. (n.d.). U.S. Department of Education, Institute of Education Sciences, National Center for Education Statistics. Available at https://nces.ed.gov/

National Center for Education Statistics. (2008). *Characteristics of private schools in the United States: Results from the 2005-2006 private school universe survey* (NCES 2008-315). Retrieved November 6, 2013, from http://nces.ed.gov/pubs2008/ pss0506/glossary.asp

National Center for Education Statistics. (2009). *Students who study science, technology, engineering, and mathematics (STEM) in postsecondary education.* Retrieved from http://nces.ed.gov/ pubs2009/2009161.pdf

National Center for Education Statistics. (2009). *Table 271: Bachelor's degrees conferred by degree-granting institutions, by field of study: Selected years, 1970-71 through 2007-08*. Retrieved from http://nces.ed.gov/programs/digest/d09/tables/ dt09_271.asp?referrer=list

National Center for Education Statistics. (2012). *The Nation's Report Card: Writing 2011* (Special Report NCES 2012–470). Institute of Education Sciences, U.S. Department of Education. Retrieved from https://nces.ed.gov/nationsreportcard/ pdf/main2011/2012470.pdf

National Center for Education Statistics. (2013). *Urban Education in America*. U.S. Department of Education. Retrieved from https://nces.ed.gov/surveys/urbaned/ students.asp

National Center for Education, Evaluation and Regional Assistance. (2016). U.S. Department of Education, Institute of Education Sciences, National Center for Education Statistics. Available at https://nces.ed.gov/

National Commission on Teahing and America's Future. (1996). *What Matters most: Teaching for America's Future*. New York: Teachers College Columbia University.

National Education Association and National Association of School Psychologists. (2007). *Truth in labeling: Disproportionality in special education*. Retrieved from http://www.nea.org/assets/docs/HE/EW-TruthInLabeling.pdf

National Education Association. (2012). *Preparing 21st century students for a global society: An educator's guide to the "Four Cs"*. Alexandria, VA: National Education Association.

National Policy Board for Educational Administration. (2015). *Professional Standards for Educational Leaders 2015*. Reston, VA: Author.

National Research Council. (2000). *How People Learn: Brain, Mind, Experience, and School*. Washington, DC: National Academy Press.

National Research Council. (2000). *Inquiry and the national science education standards: A guide for teaching and learning*. National Academies Press.

National Research Council. (2007). *Taking science to school: Learning and teaching science in grades K–8*. Washington, DC: National Academy Press. Retrieved from http://nap.edu/openbook.php?record_id=11625&page=1

National Research Council. (2011). *A framework for K–12 science education: Practices, crosscutting concepts, and core ideas*. Washington, DC: National Academy Press. Retrieved from http://nap.edu/openbook.php?record_id=13165&page=1

National Research Council. (2011). *Successful K-12 STEM education: Identifying effective approaches in science, technology, engineering, and mathematics*. National Academies Press.

National Research Council. (2011). *Successful K-12 STEM Education: Identifying Effective Approaches in Science, Technology, Engineering, and Mathematics.* Washington, DC: The National Academies Press. Retrieved from http://www.stemreports.com/wp-content/uploads/2011/06/NRC_STEM_2.pdf

National Research Council. (2012). *A Framework for K-12 Science Education: Practices, Crosscutting Concepts and Core Ideas*. Washington, DC: National Academy Press.

National Research Council. (2012). *A framework for K-12 science education: Practices, crosscutting concepts, and core ideas*. Washington, DC: National Academies Press.

National Research Council. (2012). *A framework for K-12 science education: Practices, crosscutting concepts, and core ideas*. Washington, DC: National Academies Press. Retrieved from https://www.nap.edu/catalog/13165/a-framework-for-k-12-science-education-practices-crosscutting-concepts

National Research Council. (2013). *Next Generation Science Standards: By States for States*. Washington, DC: National Academy Press.

National Science Board. (2014). *Science and Engineering Indicators 2014*. Arlington, VA: National Science Foundation.

National Science Board. (2016). *Science and Engineering Indicators 2016*. Alexandria, VA: National Science Foundation.

National Science Foundation (NSF), Division of Science Resources Statistics. (2009). *Women, minorities, and persons with disabilities in science and engineering (Special Report NSF 09-305)*. Arlington, VA: Author.

National Science Foundation (NSF). (2015). *Women, minorities, and persons with disabilities in science and engineering: 2015* (Special Report NSF 11-309). Arlington, VA: Author. Retrieved from http://www.nsf.gov/2015/nsff15311/tables.cfm

National Science Foundation, National Center for Science and Engineering Statistics. (2012). T*able 18: Doctorate recipients, by citizenship and broad field of study: Selected years, 1981–2011*. Retrieved from http://www.nsf.gov/statistics/sed/2011/pdf/tab18.pdf

National Science Foundation, National Science Board. (2010). *Preparing the next generation of STEM innovators*: *Identifying and developing our nation's human capital* (NSB-10-33). Retrieved September 19, 2013 from http://www.nsf.gov/nsb/publications/2010/ nsb1033.pdf

National Science Foundation, National Science Board. (2010). *Science and engineering indicators*. Washington, DC: The National Science Foundation. Retrieved from http://www.nsf.gov/statistics/seind10/pdf/seind10.pdf

National Science Foundation, National Science Board. (2014). *Science and engineering indicators 2014*. Washington, DC: The National Science Foundation. Retrieved from http://www.nsf.gov/statistics/seind14/content/overview/overview.pdf

National Science Foundation. (2014). *Science & Engineering Indicators*. National Science Board. NSB 14-01.

National Science Foundation. (2017). *Women, Minorities, and Persons with Disabilities in Science and Engineering*. Author.

National Science Foundation. (2017). *Women, minorities, and persons with disabilities in science and engineering*. Retrieved from https://www.nsf.gov/statistics/2017/nsf17310/

National Science Foundation. (2017). *Women, minorities, and persons with disabilities in science and engineering*. Retrieved from https://www.nsf.gov/statistics/2017/nsf17310/

National Science Teachers Association (NSTA). (2003). *Standards for science teacher preparation*. Retrieved from http://static.nsta.org/pdfs/nstastandards2003.pdf

National Staff Development Council (US). (2001). *National Staff Development Council's Standards for Staff Development*. National Staff Development Council.

Neitz, M. I. (2011). Feminist methodologies. In M. Stausberg & S. Engler (Eds.), *The Routledge Handbook of Research Methods in the Study of Religion* (pp. 54–67). New York, NY: Routledge.

Neitz, M. I. (2013). Insiders, outsiders, advocates and apostates and the religions they study: Location and the sociology of religion. *Critical Research on Religion*, *1*(2), 129–140. doi:10.1177/2050303213490311

NGSS Lead States. (2013). *Next Generation Science Standards: For States, By States*. Washington, DC: The National Academies Press.

Noebel, D., & Edwards, C. (2002). *Thinking like a Christian: Understanding and living a biblical worldview*. Nashville, TN: B&H Publishing Group.

Norman, O., Ault, C. R., Bentz, B., & Meskimen, L. (2001). The black-white? achievement gapas a perennial challenge of urban science education: A sociocultural and historical overview with implications for research and practice. *Journal of Research in Science Teaching*, *38*(10), 1101–1114. doi:10.1002/tea.10004

Nosek, B. A., Greenwald, A. G., & Banaji, M. R. (2007). The Implicit Association Test at age 7: A methodological and conceptual review. In J. A. Bargh (Ed.), *Social Psychology and the Unconscious: The Automaticity of Higher Mental Processes* (pp. 265–292). New York, NY: Psychology Press.

Nwanekezi, A. U., & Nzokurum, J. C. (2010). Science teaching in Nigerian primary schools: The way forward. *African Journal of Education and Developmental Studies, 7*(1), 68-73.

Office of State Superintendent of Education. (2017). *2017 PARCC Results*. Office of State Superintendent of Education.

Office of the State Superintendent of Education. (2016). *2016 DC PAARC Results*. Washington, DC: Office of the State Superintendent of Education.

Office of the State Superintendent of Education. (2018). *Office of the State Superintendent of Education*. Retrieved from Office of the State Superintendent of Education: https://osse.dc.gov/page/about-osse

Ong, M. (2005). Body Projects of Young Women of Color in Physics: Intersections of Gender, Race, and Science. *Social Problems, 52*(4), 593–617. doi:10.1525p.2005.52.4.593

Ong, M., Wright, C., Espinosa, L. L., & Orfield, G. (2011). Inside the Double Bind: A Synthesis of Empirical Research on Undergraduate and Graduate Women of Color in Science, Technology, Engineering, and Mathematics. *Harvard Educational Review, 81*(Summer), 172–208. doi:10.17763/haer.81.2.t022245n7x4752v2

Onuja, J. E. (1987). *The causes of poor performance of secondary schools students in West African schools certificate Biology in Oyu L. G. A.* Ado-Ekiti University Library.

Organisation for Economic Co-operation and Development. (2018). *OECD Science, Technology and Industry Scoreboard*. Retrieved from OCED.org: http://www.oecd.org/sti/scoreboard.htm

Ormrod, J. E. (2006). *Educational psychology: Developing learners*. Upper Saddle River, NJ: Pearson Education, Inc.

Osborne, J., & Collins, S. (2001). Pupils' views of the role and value of the science curriculum: A focus group study. *International Journal of Science Education, 23*(5), 441–467. doi:10.1080/09500690010006518

Oskamp, S., & Schultz, P. W. (2005). *Attitudes and opinions*. Mahwah, NJ: Lawrence Erlbaum Associates. doi:10.4324/9781410611963

Owen, J., Tao, K., & Rodolfa, E. (2010). Microaggressions and women in short-term psychotherapy: Initial evidence. *The Counseling Psychologist, 38*(7), 923–946. doi:10.1177/0011000010376093

Ozel, M. (2012). Children's images of scientists: Does grade level make a difference? *Educational Sciences: Theory and Practice*, 3187–3198.

Parson, E. C., & Summers, G. (2004). Use of images as reflective discrepant events: Pathways for elementary teachers to reconsider practice in relation to their views of science teaching and learning. *Electronic Journal of Science Education, 9*(1). Retrieved from http://ejse.southwestern.edu/article/view/7724/5491

PCAST. (2010). *Prepare and inspire: K-12 education in science, technology, engineering, and math (stem) for America's future.* PCAST.

Pearson, W. Jr, & Miller, J. D. (2012). Pathways to an engineering career. *Peabody Journal of Education, 87*(1), 46–61. doi:10.1080/0161956X.2012.642270

Penuel, W. R., Gallagher, L. P., & Moorthy, S. (2011). Preparing teachers to design sequences of instruction in earth systems science: A comparison of three professional development programs. *American Educational Research Journal, 48*(4), 996–1025. doi:10.3102/0002831211410864

Petrides, L. A., & Guiney, S. Z. (2002). Knowledge management for school leaders: An ecological framework for thinking schools. *Teachers College Record, 104*(8), 1702–1717. doi:10.1111/1467-9620.00217

Pew Forum on Religion and Public Life. (2008). *U.S. religious landscape survey.* Washington, DC: Pew Research Center. Retrieved from: http://religions.pewforum.org/pdf/report-religious-landscape-study-full.pdf

Philips, S. U. (1983). *The invisible culture.* New York: Longman.

Picardo, J. (2016, April 8). *A textbook problem: Seven suggestions to improve the quality of published resources; Exploring the cognitive principles that make a good textbook.* Retrieved from http://www.josepicardo.com/education/a-textbook-problem-seven-suggestions-to-improve-the-quality-of-published-resources/

Pierce, D. (2016). Inside Google's plan to make VR amazing for absolutely, positively everyone. *Wired.* Retrieved from https://www.wired.com/2016/04/google-vr-clay-bavor/

Pierce, C., Carew, J., Pierce-Gonzalez, D., & Willis, D. (1978). An experiment in racism: TV commercials. In C. Pierce (Ed.), *Television and education* (pp. 62–88). Beverly Hills, CA: Sage.

Pollack, J. (2015). *Shortcut: How Analogies Reveal Connections.* Spark Innovation, and Sell Our Greatest Ideas.

Porter, A., McMaken, J., Hwang, J., & Yang, R. (n.d.). Common Core State Standards: The new U.S. intended curriculum. *Educational Researcher, 40*(3), 103-115.

Pratt, H. (2007). *Science education's "overlooked ingredient": Why the path to global competitiveness begins in elementary school*. Retrieved September 10, 2013 from http://science.nsta.org/nstaexpress/nstaexpress_2007_10_29_pratt.htm

Presentation, 2.-1. D. (2018). *Office of State Superintendent of Education*. Retrieved from https://osse.dc.gov/sites/default/files/dc/sites/osse/page_content/attachments/2018%20PARCC%20Results%20Release%20%28Aug.%2016%2C%20 2018%29.pdf

President's Council of Advisors on Science and Technology (US). (2010). *Prepare and Inspire: K-12 Education in Science, Technology, Engineering, and Math (STEM) for America's Future: Executive Report*. Executive Office of the President, President's Council of Advisors on Science and Technology.

President's Council of Advisors on Science and Technology. (2010). *Report to the President: Prepare and inspire: K-12 education in science, technology, engineering, and math (STEM) for America's future. Retrieved from.* National Academies Press. Retrieved from https://www.whitehouse.gov/sites/default/files/microsites/ostp/pcast-stemed-report.pdf

President's Council of Advisors on Science and Technology. (2012). Engage to excel: Producing one million additional college graduates with degrees in science, technology, engineering, and mathematics. Washington, DC: Author.

Priestley, M., Edwards, R., Priestley, A., & Miller, K. (2012). Teacher agency in curriculum making: Agents of change and spaces for manoeuvre. *Curriculum Inquiry, 42*(2), 191–214. doi:10.1111/j.1467-873X.2012.00588.x

Programme for International Student Assessment (PISA). (n.d.). Retrieved from www.oecd.org

Puentedura, R. R. (2014). *Frameworks for Educational Technology: SAMR, the EdTech Quintet, and the Horizon Report*. Paper presented at the International Conference on Teaching and Learning with Technology (iCTLT) Conference, Singapore.

Quinn, H., Lee, O., & Valdés, G. (2012). *Language demands and opportunities in relation to Next Generation Science Standards for English language learners: What teachers need to know*. Paper presented at the Understanding Language Conference, Stanford, CA. Retrieved from http://ell.stanford.edu/papers

Redmond, A., Thomas, J., High, K., Scott, M., Jordan, P., & Dockers, J. (2011). Enriching science and math through engineering. *School Science and Mathematics*, *111*(8), 399–408. doi:10.1111/j.1949-8594.2011.00105.x

Reeve, S., & Bell, P. (2009). Children's self-documentation and understanding of the concepts 'healthy' and 'unhealthy.'. *International Journal of Science Education*, *31*(14), 1953–1974. doi:10.1080/09500690802311146

Reid, N. (2011). Attitude research in science education. In I. Saleh & M. Khine (Eds.), *Attitude research in science education: Classic and contemporary measurements* (pp. 3–44). Charlotte, NC: Information Age Publishing Inc.

Renninger, K. A. (2009). Interest and identity development in instruction: An inductive model. *Educational Psychologist*, *44*(2), 105–118. doi:10.1080/00461520902832392

Rhoads, T. R., Walden, S. E., & Winter, B. A. (2002). Sooner elementary engineering and science (SEES) clubs-an informal education experience for elementary children. *Journal for STEM Education*, *5*(3), 47–52. Retrieved from https://ieeexplore.ieee.org/document/1157993

Rhode, G., Jenson, W. R., & Reavis, H. K. (1993). *The tough kid book: Practical classroom management strategies*. Longmont, CO: Sopris West, Inc.

Rickford, J. R. (1999). *African American vernacular English: Features, evolution, educational implications*. Oxford, UK: Blackwell.

Rickford, J. R. (2005). Using the vernacular to teach the standard. In J. D. Ramirez, T. G. Wiley, G. de Klerk, E. Lee, & W. E. Wright (Eds.), *Ebonics: The urban education debate* (2nd ed.; pp. 18–40). Clevedon, UK: Multilingual Matters. doi:10.21832/9781853597985-004

Rivera, L. M., & Schaefer, M. B. (2009). The career institute: A collaborative career development program for traditionally underserved secondary (6-12) school students. *Journal of Career Development*, *35*(4), 406–426. doi:10.1177/0894845308327737

Rogoff, B. (2003). *The cultural nature of human development*. Oxford, UK: Oxford University Press.

Rosser, S. V., & Taylor, M. Z. (2009). Why are we still worried about women in science? *Academe*, *95*(3), 7–10.

Roth, K. J., Garnier, H. E., Chen, C., Lemmens, M., Schwille, K., & Wickler, N. I. (2011). Videobased lesson analysis: Effective science PD for teacher and student learning. *Journal of Research in Science Teaching*, *48*(2), 117–148. doi:10.1002/tea.20408

Rothwell, J. (2013). The Hidden STEM Economy. Washington, DC: Metropolitan Policy Program at Brookings.

Rothwell, J. (2014). *Still Searching: Job Vacancies ad STEM Skills*. Brookings Institute.

Ruiz-Primo, M. A., Li, M., & Shavelson, R. J. (2001). *Looking into students' science notebooks: What teachers do with them? CSE: Technical Report 562*. Los Angeles, CA: Center for Research on Evaluation, Standards, and Student Testing/University of California, Los Angeles.

Salta, K., & Tzougraki, C. (2004). Attitudes toward chemistry among 11th grade students in high schools in Greece. *Science Education*, *88*(4), 535–547. doi:10.1002ce.10134

Santrock, J. W. (2007). *Adolescence*. New York, NY: McGraw-Hill Companies, Inc.

Satterthwait, D. (2010). Why are 'hands-on' science activities so effective for student learning? *Teaching Science: The Journal of Australian Science Teachers Association*, *56*(2), 7–10.

Scantlebury, K. (2014). Gender matters: Building on the past, recognizing the present, and looking toward the future. In N. G. Lederman & S. K. Abell (Eds.), *Handbook of research on science education* (pp. 187–203). New York: Routledge.

Schaeffer, C. B., & Mattis, J. S. (2012). Diversity, religiosity, and spirituality in the workplace. *Journal of Management, Spirituality & Religion*, *9*(4), 317–333. doi:10.1080/14766086.2012.742750

Schleppegrell, M. J. (2004). *The language of schooling*. Mahwah, NJ: Lawrence Erlbaum Associates.

Schneider, B., Broda, M., Judy, J., & Burkander, K. (2014). Pathways to college and STEM careers: Enhancing the high school experience. *New Directions for Youth Development*, *140*, 9–29. PMID:24474256

Scifo, J. (2018). *An investigation of assessment, equal opportunity, and educational equity in conservative evangelical schools* (Doctoral dissertation). Retrieved from Liberty University Doctoral Dissertations and Projects. (1664)

Scott, A., & Martin, A. (2014). Perceived Barriers to Higher Education in STEM among High-Achieving. *Journal of Women and Minorities in Science and Engineering*, *20*(3), 235–256. doi:10.1615/JWomenMinorScienEng.2014006999

Seymour, E., & Hewitt, N. (1994). *Talking about leaving: Factors contributing to high attrition rates among science, mathematics & engineering undergraduate majors*. Final Report to the Alfred P. Sloan Foundation. Boulder, CO: University of Colorado, Ethnography and Assessment Research Bureau of Sociological Research.

Seymour, E., & Hewitt, N. (1997). *Talking about leaving: Why undergraduates leave the sciences*. Boulder, CO: Westview.

Shapiro, C. A., & Sax, L. J. (2011). Major selection and persistence for women in STEM. *New Directions for Institutional Research*, *152*(152), 5–18. doi:10.1002/ir.404

Shaw, E. J., & Barbuti, S. (2010). Patterns of persistence in intended college major with a focus on STEM majors. *NACADA Journal*, *30*(2), 19–34. doi:10.12930/0271-9517-30.2.19

Shenouda, C. K., & Danovitch, J. H. (2014). Effects of gender stereotypes and stereotype threat on children's performance on a spatial task. *Revue Internationale de Psychologie Sociale*, *27*(3–4), 53–77.

Shepardson, D. P., & Britsch, S. J. (2001). The role of children's journals in elementary school science activities. *Journal of Research in Science Teaching*, *38*(1), 43–69. doi:10.1002/1098-2736(200101)38:1<43::AID-TEA4>3.0.CO;2-I

Shifrer, D., Muller, C., & Callahan, R. (2011). Disproportionality and learning disabilities: Parsing apart race, socioeconomic status, and language. *Journal of Learning Disabilities*, *44*(3), 246–257. doi:10.1177/0022219410374236 PMID:20587753

Shoffner, M. F., Newsome, D., Barrio Minton, C. A., & Wachter Morris, C. A. (2015). A qualitative exploration of the STEM career-related outcome expectations of young adolescents. *Journal of Career Development*, *42*(2), 102–116. doi:10.1177/0894845314544033

Sjøberg, S., & Schreiner, C. (2007). Perceptions and images of science and science education. In M. Claessens (Ed.), *Communicating European Research 2005* (pp. 151–158). Springer Netherlands. doi:10.1007/1-4020-5358-4_26

Sleeter, C. E. (2008). Preparing White teachers for diverse students. In M. Cochran-Smith, S. Feiman-Nemser, & J. McIntyer (Eds.), *Handbook of research in teacher education: Enduring issues in changing contexts* (3rd ed.; pp. 559–582). New York, NY: Routledge.

Sleeter, C. E., & Owuor, J. (2011). Research on the impact of teacher preparation to teach diverse students: The research we have and the research we need. *Action in Teacher Education*, *33*(5-6), 524–536. doi:10.1080/01626620.2011.627045

Smith, J., & Karr-Kidwell, P. (2000). *The interdisciplinary curriculum: A literary review and a manual for administrators and teachers*. Retrieved from ERIC database. (ED443172)

Smith, E. (2010). Do we need more scientists? A long-term view of patterns of participation in UK undergraduate science programmes. *Cambridge Journal of Education*, *40*(3), 281–298. doi:10.1080/0305764X.2010.502886

Smith, E. (2011). Women into science and engineering? Gendered participation in higher education STEM subjects. *British Educational Research Journal*, *37*(6), 993–1014. doi:10.1080/01411926.2010.515019

Smith, J. L., Sansone, C., & White, P. H. (2007). The stereotyped task engagement process: The role of interest and achievement motivation. *Journal of Educational Psychology*, *99*(1), 99–114. doi:10.1037/0022-0663.99.1.99

Solorzano, D. G. (1998). Critical race theory, race and gender microaggressions, and the experience of Chicana and Chicano scholars. *International Journal of Qualitative Studies in Education: QSE*, *11*(1), 121–136. doi:10.1080/095183998236926

Sondergeld, T. A., Johnson, C. C., & Walten, J. B. (2016). Assessing the Impact of a Statewide STEM Investment on K-12, Higher Education, and Business/Community STEM Awareness Over Time. *School Science and Mathematics*, *116*(2), 104–110. doi:10.1111sm.12155

Sonnert, G., Sadler, P., & Michaels, M. (2013). Gender Aspects of Participation, Support, and Success in a State Science Fair. *School Science and Mathematics*, *113*(3), 135–143. doi:10.1111sm.12007

Sparks, D. (2016). Reducing stereotype threat in the science and mathematics classroom: An overview of research, best practices, and intervention strategies. *Currents in Teaching and Learning*, *7*(2), 4–17.

Spearman, J., & Watt, H. M. (2013). Women's aspirations towards STEM careers. In W. Patton (Ed.), *Conceptualising Women's Working Lives* (pp. 175–191). Boston, MA: SensePublishers. doi:10.1007/978-94-6209-209-9_10

Standards in Your State. (2013). Retrieved from Common Core State Standards Initiative: http://www.corestandards.org/standards-in-your-state

Steele, C. M. (1997). A threat in the air: How stereotypes shape intellectual identity and performance. *The American Psychologist*, *52*(6), 613–629. doi:10.1037/0003-066X.52.6.613 PMID:9174398

Steele, C. M., & Aronson, J. (1995). Stereotype threat and the intellectual test performance of African Americans. *Journal of Personality and Social Psychology*, *69*(5), 797–811. doi:10.1037/0022-3514.69.5.797 PMID:7473032

Steele, D. M., & Cohn-Vargas, B. (2013). *Identity safe classrooms: Places to belong and learn*. Thousand Oaks, CA: Corwin Press.

Stoeger, H., Duan, X., Schirner, S., Greindl, T., & Ziegler, A. (2013). The effectiveness of a one-year online mentoring program for girls in STEM. *Computers & Education*, *69*, 408–418. doi:10.1016/j.compedu.2013.07.032

Stoll, L., Bolam, R., McMahon, A., Wallace, M., & Thomas, S. (2006). Professional learning communities: A review of the literature. *Journal of Educational Change*, *7*(4), 221–258. doi:10.100710833-006-0001-8

Stuessy, N. E. (2015). Modeling Successful STEM High Schools in the United States: An Ecology Framework. *International Journal of Education in Mathematics, Science and Technology*, *3*(1), 77–92.

Subotnik, R., Tai, R., Rickoff, R., & Almarode, J. (2009). Specialized Public High Schools of Science, Mathematics, and Technology and the STEM Pipeline: What Do We Know Now and What Will We Know in 5 Years? *Roeper Review*, *32*(1), 7–16. doi:10.1080/02783190903386553

Sue, D. W. (2010). *Microaggressions in everyday life: Race, gender, and sexual orientation*. Hoboken, NJ: Wiley.

Sue, D. W., Bucceri, J., Kin, A. I., Nadal, K. L., & Torino, G. C. (2007). Racial microaggressions and the Asian American experience. *Cultural Diversity & Ethnic Minority Psychology*, *13*(1), 72–81. doi:10.1037/1099-9809.13.1.72 PMID:17227179

Sue, D. W., Lin, A. I., Torino, G. C., Capodilupo, C. M., & Rivera, D. P. (2009). Racial microaggressions and difficult dialogues on race in the classroom. *Cultural Diversity & Ethnic Minority Psychology*, *15*(2), 183–190. doi:10.1037/a0014191 PMID:19364205

Sullivan, A. L. (2011). Disproportionality in special education identification and placement of English language learners. *Exceptional Children*, *77*(3), 317–334. doi:10.1177/001440291107700304

Supovitz, J. A., & Turner, H. M. (2000). The effects of professional development on science teaching practices and classroom culture. *Journal of Research in Science Teaching*, *37*(9), 963–980. doi:10.1002/1098-2736(200011)37:9<963::AID-TEA6>3.0.CO;2-0

Su, R., Rounds, J., & Armstrong, P. I. (2009). Men and things, women and people: A meta-analysis of sex differences in interests. *Psychological Bulletin*, *135*(6), 859–884. doi:10.1037/a0017364 PMID:19883140

Suri, G., & Sharma, S. (2013). The impact of gender on attitude towards computer technology and e-learning: An exploratory study of Punjab University, India. *International Journal of Engine Research*, *2*(2), 132–136.

Swanson, J. L., & Fouad, N. A. (2014). *Career theory and practice: Learning through case studies* (3rd ed.). Thousand Oaks, CA: Sage.

Syed, M., Goza, B. K., Chemers, M. M., & Zurbriggen, E. L. (2012). Individual differences in preferences for matched-ethnic mentors among high-achieving ethnically diverse adolescents in STEM. *Child Development*, *83*(3), 896–910. doi:10.1111/j.1467-8624.2012.01744.x PMID:22506789

Taraban, R., Box, C., Myers, R., Pollard, R., & Bowen, C. (2007). Effects of active-learning experiences on achievement, attitudes, and behaviors in high school biology. *Journal of Research in Science Teaching*, *44*(7), 960–979. doi:10.1002/tea.20183

Tellhed, U., Bäckström, M., & Björklund, F. (2017). Will I fit in and do well? The importance of social belongingness and self-efficacy for explaining gender differences in interest in STEM and HEED majors. *Sex Roles*, *77*(1), 86–96. doi:10.100711199-016-0694-y PMID:28725103

The After School Alliance. (2011). *STEM Learning in Afterschool: An Analysis of Impact and Outcomes*. Retrieved from http://www.afterschoolalliance.org/stem-afterschool-outcomes.pdf

The After School Alliance. (2013). *Defining Youth Outcomes for STEM Learning in Afterschool*. Author.

The Condition of Educaton: Concentration of Public School Students Eligible for Free or Reduced-Price Lunch. (2018). Institute of Education Sciences: National Center for Education Statistics.

Thompson, M. T., Marchant, M., Anderson, D., Prater, M. A., & Gibb, G. (2012). Effects of tiered training on general educators' use of specific praise. *Education & Treatment of Children*, *35*(4), 521–546. doi:10.1353/etc.2012.0032

Thompson, S. C., Gregg, L., & Niska, J. M. (2004). Professional learning communities, leadership, and student learning. *RMLE Online: Research in Middle Level Education*, *28*(1), 15. doi:10.1080/19404476.2004.11658173

Thurston, A., Topping, K., Tolmie, A., Christie, D., Karagiannidou, E., & Murray, P. (2010). Cooperative learning in science: Follow-up from primary to high school. *International Journal of Science Education*, *32*(4), 501–522. doi:10.1080/09500690902721673

Tienken, C. (2011). *CCCS: I wonder? Kappa Delta Pi Record*. Retrieved from ERIC database. (EJ939062)

Tirozzi, G., & Uro, G. (1997). Education reform in the United States: National policy in. *The American Psychologist*, *52*(3), 241–249. doi:10.1037/0003-066X.52.3.241

Tooze, A. (2008). *The Wages of Destruction: The Making and Breaking of the Nazi Economy*. Penguin Books.

Torres, L., Driscoll, M. W., & Burrow, A. L. (2010). Racial microaggressions and psychological functioning among highly achieving African-Americans: A mixed methods approach. *Journal of Social and Clinical Psychology*, *29*(10), 1074–1099. doi:10.1521/jscp.2010.29.10.1074

Tseng, K. H., Chang, C. C., Lou, S. J., & Chen, W. P. (2013). Attitudes towards science, technology, engineering and mathematics (STEM) in a project-based learning (PjBL) environment. *International Journal of Technology and Design Education*, *23*(1), 87–102. doi:10.100710798-011-9160-x

Tuan, H.-L., Chin, C.-C., & Shieh, S.-H. (2005a). The development of a questionnaire to measure students' motivation towards science learning. *International Journal of Science Education*, *27*(6), 639–654. doi:10.1080/0950069042000323737

Tucker-Raymond, E., Varelas, M., Pappas, C., Korzh, A., & Wentland, A. (2007). "They probably aren't named Rachel": Young children's scientist identities as emergent multimodal narratives. *Cultural Studies of Science Education*, *1*(3), 559–592. doi:10.100711422-006-9017-x

Türkmen, H. (2008). Turkish primary students' perceptions about scientists and what factors affecting the image of scientists. *Eurasia Journal of Mathematics, Science &Technology Education*, *4*(1), 55–61.

Turner, S., & Ireson, G. (2010). Fifteen pupils' positive approach to primary school science: When does it decline? *Educational Studies*, *36*(2), 119–141. doi:10.1080/03055690903148662

U.S. Census Bureau. (2012, April 8). *Recent population trends for the U.S. island areas: 2000 to 2010* (Report No. P23-213). Washington, DC: U.S. Government Printing Office.

U.S. Census Bureau. (2014, July 10). *Where do college graduates work?* Retrieved from United States Census Bureau: https://www.census.gov/dataviz/visualizations/stem/stem-html/

U.S. Department of Education. (2009). *Race to the Top Program Executive Summary*. Washington, DC: U.S. Department of Education.

U.S. Department of Education. (2016). *STEM 2026 A Vision for Innovation in STEM Education*. U.S. Department of Education.

U.S. Department of Education. (2016). *Summer Bridge Programs*. What Works Clearinghouse.

U.S. News/Raytheon STEM INdex. (2016). STEM Index. *U.S. News & World Report*.

United States Department of Labor (USDOL). (2013). *Employment projections: 2012-2022 summary*. Washington, DC: Author. Retrieved from http://www.bls.gov/news.release/ecopro.nr0.htm

United States Government Accountability Office. (2012). *Science, Technology, Engineering, and Mathematics Education: Strategic Planning Needed to Better Manage Overlapping Programs Across Multiple Agencies*. United States Government Accountability Office.

University of the District of Columbia. (n.d.). *The Equity Imperative: A Strategic Plan to Regenerate the University of the District of Columbia As A Public Higher Education Model of Urban Student Success.* Washington, DC: University of the District of Columbia.

Urban Education in America. (2013). Institute of Education Science: National Center Education Statistics.

Usher, E. L., & Pajares, F. (2009). Sources of self-efficacy in mathematics: A validation study. *Contemporary Educational Psychology*, *34*(1), 89–101. doi:10.1016/j.cedpsych.2008.09.002

Valdés, G. (2004). Between support and marginalisation: The development of academic language in linguistic minority children. *International Journal of Bilingual Education and Bilingualism*, *7*(2-3), 102–132. doi:10.1080/13670050408667804

Valencia, R. R. (1997). Conceptualizing the notion of deficit thinking. In R. R. Valencia (Ed.), *The evolution of deficit thinking: Educational thought and practice* (pp. 1–12). Bristol, PA: Falmer Press.

Van Driel, J. H., & Berry, A. (2012). Teacher professional development focusing on pedagogical content knowledge. *Educational Researcher*, *41*(1), 26–28. doi:10.3102/0013189X11431010

van Lier, L., & Walqui, A. (2012). *Language and the Common Core State Standards.* Paper presented at the Understanding Language Conference, Stanford, CA. Retrieved from http://ell.stanford.edu/papers

Varelas, M., Kane, J. M., & Wylie, C. D. (2011). Young African American children's representations of self, science, and school: Making sense of difference. *Science Education*, *95*(5), 824–851. doi:10.1002ce.20447

Vendetti, M. S., Matlen, B. J., Richland, L. E., & Bunge, S. A. (2015). Analogical reasoning in the classroom: Insights from cognitive science. *Mind, Brain and Education: the Official Journal of the International Mind, Brain, and Education Society*, *9*(2), 100–106. doi:10.1111/mbe.12080

Wadsworth, N. D. (2011). Intersectionality in California's same-sex marriage battles: A complex proposition. *Political Research Quarterly*, *64*(1), 200–216. doi:10.1177/1065912910376386

Waitoller, F. R., Artiles, A. J., & Cheney, D. A. (2010). The miner's canary a review of overrepresentation research and explanations. *The Journal of Special Education, 44*(1), 29–49. doi:10.1177/0022466908329226

Wang, M. C., & Kovach, J. A. (1995). *Bridging the Achievement Gap in Urban Schools: Reducing Educational Segregation and Advancing Resilience--Promoting Strategies*. Publication Series# 95-9.

Wang, H., & Billington, B. (2016). Economically disadvantaged minority girls' knowledge and perceptions of science and engineering and related careers. *Journal of Extension, 54*(6), 1–8.

Wang, J., Leu, J., & Shoda, Y. (2011). When the seemingly innocuous "stings": Racial microaggressions and their emotional consequences. *Personality and Social Psychology Bulletin, 37*(12), 1666–1678. doi:10.1177/0146167211416130 PMID:21885859

Ware, J., & Stein, S. (2013). From "Mentor" to "Role Model": Scaling the involvement of STEM professionals through role model videos. *Journal of Educational Multimedia and Hypermedia, 22*(2), 209–223.

Warner, L. R., & Shields, S. A. (2013). The intersections of sexuality, gender, and race: Identity research at the crossroads. *Sex Roles, 68*(11-12), 803–810. doi:10.100711199-013-0281-4

Warren, B., Ballenger, C., Ogonowski, M., Rosebery, A. S., & Hudicourt-Barnes, J. (2001). Rethinking diversity in learning science: The logic of everyday sense-making. *Journal of Research in Science Teaching, 38*(5), 529–552. doi:10.1002/tea.1017

Watson, D. (2011). What do you mean when you say Urban? *Rethinking Schools, 26*(1), 48–50.

Watt, H. M. G. (2008). Gender and occupational outcomes: An introduction. In H. M. G. Watt & J. S. Eccles (Eds.), *Gender and occupational outcomes: Longitudinal assessments of individual, social, and cultural influences* (pp. 3–24). Washington, DC: American Psychological Association. doi:10.1037/11706-000

Weinbaum, A., Allen, D., Blythe, T., Simon, K., Seidel, S., & Rubin, C. (2004). *Teaching as inquiry*. New York, NY: Teachers College Press.

Wei, R., Darling-Hammond, L., Andree, A., Richardson, N., & Orphanos, S. (2009). *Professional learning in the learning profession: A status report on teacher*. Dallas, TX: National Staff.

Weisgram, E. S., & Bigler, R. S. (2007). Effects of learning about gender discrimination on adolescent girls' attitudes toward and interest in science. *Psychology of Women Quarterly, 31*(3), 262–269. doi:10.1111/j.1471-6402.2007.00369.x

Weis, L., Eisenhart, M., Cipollone, K., Stich, A. E., Nikischer, A. B., Hanson, J., ... Dominguez, R. (2015). In the Guise of STEM Education Reform: Opportunity Structures and Outcomes in Inclusive STEM-Focused High Schools. *American Educational Research Journal, 52*(6), 1024–1059. doi:10.3102/0002831215604045

Welch, A., & Huffman, D. (2011). The effect of robotics competitions on high school students' attitudes toward science. *School Science and Mathematics, 111*(8), 416–424. doi:10.1111/j.1949-8594.2011.00107.x

Wendt, J. L., Rockinson-Szapkiw, A. J., & Cordes, M. (2017). Examining the influence of a STEM certification model on female, minority science outcomes. *The Journal of Educational Research, 111*(6), 666–677. doi:10.1080/00220671.2017.1396437

Will, M. (2018). *'An Expensive Experiment': Gates Teacher-Effectiveness Program Shows No Gains for Students.* Retrieved from https://www.edweek.org/ew/articles/2018/06/21/an-expensive-experiment-gates-teacher-effectiveness-program-show.html

Williams, B. (Ed.). (1996). *Closing the achievement gap: A vision for changing beliefs and practices.* ASCD.

Wilson, L. (n.d.). *The Second Principle - The Work of Leslie Wilson, Ed. D.* Retrieved from The Second Principle: https://thesecondprinciple.com/teaching-essentials/beyond-bloom-cognitive-taxonomy-revised/

Wissehr, C., Concannon, J., & Barrow, L. H. (2011). Looking back at the Sputnik era and its impact on science education. *School Science and Mathematics, 111*(7), 368–375. doi:10.1111/j.1949-8594.2011.00099.x

Wist, C. (2006). *Putting it all together: Understanding the research behind interactive notebooks.* Retrieved from https://tccl.arcc.albany.edu/knilt/images/d/d0/Interactive_Notebooks_Research.pdf

Wong Fillmore, L., & Snow, C. (2018). What teachers need to know about language. In C. T. Adger, C. E. Snow, & D. Christian (Eds.), *What Teachers Need to Know about Language* (2nd ed.). Bristol, UK: Multilingual Matters.

Wood, L., & Howley, A. (2012). Dividing at an early age: The hidden digital divide in Ohio elementary schools. *Learning, Media and Technology*, *37*(1), 20–39. doi:10.1080/17439884.2011.567991

World Economic Forum. (2009). *The Global Competitiveness Report 2009-2010.* Retrieved from http://www.weforum.org/pdf/GCR09/Report/Countries/United%20States.pdf

Wyss, V. L., Heulskamp, D., & Seibert, C. J. (2012). Increasing middle school student interest in STEM careers with videos of scientists. *International Journal of Environmental and Science Education*, *7*(4), 501–522.

Xie, Y., & Shauman, K. (2015). STEM Education. *Annual Review of Sociology*, *41*(1), 331–357. doi:10.1146/annurev-soc-071312-145659 PMID:26778893

Yin, R. K. (2009). *Case study research: Design and methods* (4th ed.). Thousand Oaks, CA: Sage Publications.

Yoon, K. S., Duncan, T. W.-Y., Lee, S., Scarloss, B., & Shapley, K. L. (2007). *Reviewing the evidence on how teacher professional development affects student achievement.* U.S. Department of Education, Institute of Education Sciences. National Center for Education Evaluation and Regional Assistance, Regional Educational Laboratory Southwest. Retrieved from http://ies.ed.gov/ncee/edlabs/regions/southwest/pdf/rel_2007033_sum.pdf

Young, J. L., Feille, K. K., & Young, J. R. (2017). Black girls as learners and doers of science: A single-group summary of elementary science achievement. *Electronic Journal of Science Education, 21*(2).

Zantua, L. (2017). Utilization of virtual reality content in grade 6 social studies using affordable virtual reality technology. *Asia Pacific Journal of Multidisciplinary Research, 5*(2), 1-10. Retrieved from https://doaj.org/article/2fe23fbd126948b08cffc832a5018a7

Zepeda, S. J. (2012). *Professional development: What works*. Larchmont, NY: Eye on Education.

Zepeda, S. J., Parylo, O., & Bengtson, E. (2014). Analyzing principal professional development practices through the lens of adult learning theory. *Professional Development in Education*, *40*(2), 295–315. doi:10.1080/19415257.2013.821667

Zhai, J., Jocz, J., & Tan, A. (2013). 'Am I like a scientist?': Primary children's images of doing science in school. *International Journal of Science Education*, *36*(4), 553–576. doi:10.1080/09500693.2013.791958

Zhao, Y., & Frank, K. A. (2003). Factors affecting technology uses in schools: An ecological perspective. *American Educational Research Journal*, *40*(4), 807–840. doi:10.3102/00028312040004807

Zucker, A. A., Shields, P. M., & Adelman, N. E. (1998). *A Report on the Evaluation fo the National Science Foundations's Statewide Systemic Initiatives (SSI)*. Program.

Zuger, S. (2012). The state of STEM (science, technology, engineering, and mathematics). *Technology and Learning*, *33*(3), 38–44.

About the Contributors

Jillian L. Wendt has worked in the field of education as a K-12 science educator, curriculum specialist, and university professor at the undergraduate, graduate, and doctoral levels for over a decade. Currently, she serves as Assistant Professor of Science Education in the Division of Education, Health, and Social Work, College of Arts & Sciences at the University of the District of Columbia (UDC) in Washington, DC. She first studied at Virginia Commonwealth University where she received her Bachelors of Science in Biology. After discovering her passion as an educator, she earned a Masters of Education in Teaching and Learning, Education Specialist in Teaching and Learning, and Doctor of Education in Curriculum and Instruction from Liberty University. In addition to researching, leading grant projects, and teaching at UDC, Dr. Wendt serves an active role in the broader education community. You will often find her volunteering for professional organizations, providing professional development opportunities, and partnering with education boards and institutions locally and nationally. Dr. Wendt's work has been recognized through receipt of the Association of Teacher Educators Robert Stevenson Scholarship, the Scholar of Promise Award for the College of Arts and Sciences at UDC, and the Accomplishment in Research/Scholarship Award for the College of Arts and Sciences at UDC, among other accolades. Dr. Wendt's areas of expertise and research interests include technology in education, science education, STEM education, and teacher preparation.

Danielle Apugo is an Assistant Professor of Urban Education at the University of the District of Columbia (UDC). Her research broadly examines race, gender, and culture dynamics in Urban Education.

* * *

Philip Alsup is the Assistant Vice President of Online and Distance Learning at Maranatha Baptist University in Watertown, WI, where he also teaches Educational Psychology and Action Research. His background and experience include K-12 teaching and administration in Christian education as well as online instruction at the graduate and doctoral levels. His research and publishing background include STEM education, classroom planning, and curriculum design. He holds a B.A. in PreLaw, an M.S. in Educational Leadership, and an Ed.D. in Curriculum and Instruction.

Seidah Armstrong is a doctoral candidate at Virginia Polytechnic and State University. With over 20 years of experience in PreK-12 education, she has worked in urban, rural, and suburban educational environments. As a former middle school principal, she enhanced the schools she touched through STEM education. Throughout her career she has forged partnerships to create profound experiences for children in education. Seidah is the author of the children's book ABC's STEM & Me. She is currently the Vice President of Performance Management Services, in which she heads the education department focusing on curriculum development, educational consultant, and professional development, which includes financial literacy and workforce development for adults. Seidah also contributed work to Tablet Computers in School Libraries and Classrooms, published in 2014. Seidah is a member of Alpha Kappa Alpha Sorority Inc. When she is not travelling, she volunteers her time introducing the world of STEM to youth of all ages. Seidah resides in northern Virginia with her husband John.

Logan Caldwell graduated from the University of North Carolina at Chapel Hill in 2003 with a BA in Elementary Education (K-6). She taught 2nd, 3rd, and 4th grades in public schools for 13 years in North Carolina and Tennessee. She achieved her initial National Board Certification as a Middle Childhood Generalist in 2007 and successfully renewed in 2017. She earned her Master's in Instruction and Curriculum Leadership in Instructional Design and Technology from the University of Memphis in August 2011 and is presently pursuing her doctoral degree in the same field. She is currently serving as a teacher educator and site coordinator in the elementary education department at the University of Memphis. She has also earned her Instructional Computing Applications certificate in 2012 and her School Library Information Specialist (K-12) endorsement in 2014 from the University of Memphis.

Annie Duguay is the Director of Language and Literacy at the Center for Applied Linguistics (CAL) in Washington, D.C. Annie coordinates and delivers professional development for educators working with linguistically diverse students in school districts throughout the country and helps schools to identify their needs for professional development to improve educational outcomes for language learners.

Additionally, Annie provides technical assistance on projects developing curricula and advises on second language literacy projects internationally. Annie has also been involved with several vocabulary intervention project, including the CREATE projects (Center for Research on the Educational Achievement and Teaching of English Language Learners) and the NICHD-funded Vocabulary Instruction and Assessment of Spanish-speakers (VIAS) project. Prior to coming to CAL, Annie was a certified ESL teacher and taught at a STEM-focused public charter school in Cambridge, Massachusetts. Annie holds two master's degrees: an Ed.M. from the Harvard Graduate School of Education and an M.A. in Immigration and Settlement Studies from Ryerson University in Toronto.

Shurron Farmer is currently an assistant professor of mathematics at the University of the District of Columbia. He received his B.S. in mathematics in 1994 from Florida A&M University in Tallahassee Florida, his M.S. in mathematics in 1996 from Howard University in Washington, DC, and his Ph.D. in mathematics in 2001 from Howard University. He is originally from Quincy, Florida and is the youngest of four children. Dr. Farmer's mathematical research interests are primarily in the areas of educational technology and mathematics higher education. He completed his doctoral dissertation in difference equations and mathematical biology, and continues to explore this area of mathematics among others. Dr. Farmer is highly interested in increasing the number of persons of color who pursue Master's and Doctoral degrees in STEM-related fields. He would like to express the interest at the level of the professorate through participation in programs designed to increase the number of persons of color who pursue advanced degrees in the areas listed above, and by contributing to the advancement of the mathematics department of which he is a member.

Jeffery Fleming is an Associate Professor of Mathematics and Associate Dean of UDC's College of Arts and Sciences. Dr. Fleming has established himself as a student-centered advocate for STEM education in formal and informal learning environments. Dr. Fleming is an energetic and innovative educator with over 15 years of teaching experience at all levels of undergraduate mathematics to diverse populations of students including first-generation students, transfer students, and adult learners. Dr. Fleming continues to show his commitment to underrepresented and underprepared students, especially students in the STEM pipeline, by providing high quality educational experiences that use high impact practices that meet students where they are and shift them to greater understanding of mathematical content by bridging the gap between mathematical theory and mathematical application.

Ashley Huderson is a native of New Orleans, LA, and completed her undergraduate training at Spelman College (2006), a certificate in Health Policy (2012) and doctoral work at Meharry Medical College (2013). A culmination of personal and academic interest in health policy, prompted her to seek out and accept a postdoctoral fellowship position at Georgetown University Lombardi Cancer Center's Office of Health Disparities and Minority Research (2015). During her two years at Georgetown University her interest in exposing and helping minority students navigate their STEM careers flourished as she accepted her first adjunct position, affording her the opportunity to teach and advise undergraduate and graduate level students. Serving as an instructor and researcher, exposed her to a number of well-established and emerging educational practices that related to fostering students' academic achievements, interest, and professional development. It was during this time that she decided to turn her sights completely to diversity and inclusion issues within STEM education and embark on a career that would allow her to make a meaningful contribution on diversifying the scientific workforce and empowering those interested in STEM, regardless of their background. Dr. Huderson was a 2015-2017 American Association for the Advancement of Science, Science and Technology Policy (AAAS S&T) Fellow in the Engineering Education and Centers' division (EEC) at the National Science Foundation, where she provided leadership on developing, coordinating, and implementing support for programs that foster an inclusive climate for pre-collegiate and collegiate STEM students. Currently Dr. Huderson serves as the Manager of Engineering Education at the American Society of Mechanical Engineers (ASME), where she is responsible for advancing and managing the research, development, promotion, implementation, and assessment of products and services that will help colleges of engineering develop their curricula and faculty. She also manages all ASME/ABET operational activities, annual program evaluator selection and assignment to ABET accreditation visits, and ASME evaluator recruitment, selection, training, and evaluation. In addition, she is also the founder of STEM Innovation Consulting, a consulting firm providing STEM educational consulting services that include professional development, project development and management, curriculum development, grant writing, and stakeholder networking to build strategic partnerships between industry, educators, researchers, and community activists. Dr. Huderson has extensive experience as a skilled facilitator, leader, academic trainer, and STEM (Science, Technology, Engineering, Arts and Mathematics) curriculum designer and advocate. Dr. Huderson has over 12 years of STEM outreach and advocacy in the community developing informal science programs, mentoring STEM majors, equipping parents with tools and resources to encourage their children in STEM, and providing culturally relevant science education pedagogy to teachers. Dr. Huderson also works with local and Federal government agencies, non-profits, and academic institutions to provided leadership support for programs that foster an inclusive climate for pre-collegiate and collegiate STEM students.

Brandy Huderson is a New Orleans native, now residing in Washington, DC. Dr. Huderson completed her undergraduate studies at Xavier University of Louisiana (New Orleans, LA) receiving a bachelor's degree in Biology, with a focus on Pre-Med. During her junior of college, Dr. Huderson decided that medical was not the right fit and instead enter the Biological Sciences Graduate program at the University of New Orleans (LA). She earned a Master's of Biological Sciences degree with concentrations in Molecular Genetics. Two years after completing her MS degree, Dr. Huderson joined the Dairy Science Department at Virginia Tech (Blacksburg, VA) where dissertation research focused on mammary gland development in pre-pubertal Holstein calves. Dr. Huderson completed two postdoctoral fellowships at Tulane University School of Medicine (New Orleans, LA) and Georgetown University's Lombardi Comprehensive Cancer Center (Washington, DC). Dr. Huderson is currently an Assistant Professor at the University of the District of Columbia in our nation's capital. Dr. Huderson's research program focuses on steroid receptor biology, in diseased and abhorrent systems, as well as STEM Education. Dr. Huderson is the oldest of two girls, her sister is "the other" Dr. Huderson. This Dr. Huderson currently resides in Washington, DC with her husband and fur baby, Nyah.

Jennifer Renn (PhD, University of North Carolina at Chapel Hill) is a Senior Project Manager in the College of Education at Purdue University. Prior to joining Purdue, she was the Director of Linguistic and Cultural Diversity at the Center for Applied Linguistics in Washington D.C., where she directed the English for Heritage Language Speakers Program and worked in K-12 English language assessment research. Dr. Renn was an Institute of Education Sciences Postdoctoral Fellow in Early Childhood Education at the Frank Porter Graham Child Development Institute at the University of North Carolina at Chapel Hill. During her time there, she worked on various projects investigating the relationship between language use and academic achievement outcomes in minority youth and contributed to writing grants for the National Science Foundation and the Institute of Education Sciences.

Amanda Rockinson-Szapkiw holds an EdD in Distance Education and a PhD in Counselor Education and Supervision. She is an Associate Professor for the Instructional Design and Technology (IDT) Program at the University of Memphis (UofM). She serves as coordinator of the IDT EdD program. Prior to her appointment at UofM, she served as an Associate Professor and program chair of Research and Doctoral programs for a private institution in Central Virginia. Dr. Rockinson-Szapkiw has authored and co-authored more than four dozen peer-reviewed journal articles and presented over 50 presentations, primarily focusing on technology integration in the distance education, doctoral persistence, and the advancement of women in higher education. She co-edited with Lucinda Spaulding the book, Navigating the Doctoral Journey: A Handbook of Strategies for Success, which received a 2015 AERA SIG 168 Outstanding Publication Award. Her development of a collaborative workspace to facilitate doctoral mentorship in the online environment was recognized by Microsoft via a case study and was awarded the 2013 Campus Technology innovator award.

Katherine Wade-Jaimes is an Assistant Professor of Education at the University of Memphis where she teaches courses in science education. Her research explores diversity and equity in science education, including critical evaluations of the societal forces that lead to marginalization from science.

Otto Wilson Jr. is an Associate Professor in the Department of Biomedical Engineering at the Catholic University of America.

Zora Wolfe is an Assistant Professor of Education, at Widener University (Chester, PA) where she primarily teaches in the K-12 Educational Leadership Program. Her areas of expertise include teacher leadership, collaborative inquiry communities, and developing teacher practice, particularly around math, science and interdisciplinary teaching. Her current research focuses on the role of principals in supporting the development of teacher leaders to help implement sustained school reform.

Index

O

P

R

S

T

U

www.ingramcontent.com/pod-product-compliance
Lightning Source LLC
LaVergne TN
LVHW081338300926
849161LV00008B/2007
* 9 7 8 1 5 2 2 5 7 8 1 4 7 *